Michael Seidel

Tensile Surface Structures

A Practical Guide to Cable
and Membrane Construction

Tensile Surface Structures
A Practical Guide to Cable and Membrane Construction

Michael Seidel

Materials

Design

Assembly and Erection

Dipl.-Ing. Dr. Michael Seidel
Institut für Architektur und Entwerfen
Abteilung Hochbau 2 – Konstruktion, Installation
und Entwerfen
Technische Universität Wien
Karlsplatz 13, 1040 Wien, Austria

Cover: detail drawing of the Sony-Center panel roof, Berlin, Germany (Waagner-Biro, Wien, Austria)

This book contains 371 illustrations.

Bibliographic information published by the
Deutsche Nationalbibliothek
The Deutsche Nationalbibliothek lists this publication in the Deutsche Nationalbibliographie; detailed bibliographic data is available in the Internet at <http://dnb.ddb.de>.

ISBN 978-3-433-02922-0
© 2009 Ernst & Sohn Verlag für Architektur und technische Wissenschaften GmbH & Co. KG, Berlin, Germany

All rights reserved (including those of translation into other languages). No part of this book may be reproduced in any form – by photoprinting, or any other means – nor transmitted or translated into a machine language without written permission from the publisher. Registered names, trademarks, etc. used in this book, even when not specifically marked as such, are not to be considered unprotected by law.

English translation: David Sturge, Kirchbach, Germany
Layout and production: 29 transfer concept
Printing: betz-druck GmbH, Darmstadt, Germany
Binding: Litges & Dopf Buchbinderei, Heppenheim, Germany
Printed in Germany

Foreword

We have here a remarkable book. It represents the first exhaustive description produced on a scientific basis of all the knowledge and interactions required for construction with flexible structural elements, covering ropes and belts, fabrics, yarns and foils.

The word "construction" can be taken literally: the author Michael Seidel considers the processes of production and erection of tensile surface structures, their interactions with the design work of architects and engineers and also the quality of the final end-product "structure". He advances here into an area, which almost all specialists in lightweight construction recognise, but where they have no overview of the details and are clearly not proficient. The results of this lack can be seen almost daily: either the erection as intended is impractical or even impossible, and completed design work has to be altered or sometimes thoroughly reworked; perhaps even more seriously, the end-product "structure" cannot be built in the intended quality.

It has long been overdue that the manufacturing and assembly processes in construction with flexible structural elements be collected, categorised and evaluated. Not only the overview produced here is of great value to all those concerned with the construction of tensile surface structures, rather it is the presentation of the interactions between design, erection and the quality of the finished product, which makes the present book so important in filling a wide gap in the existing literature on lightweight construction.

Stuttgart, October 2007 *Werner Sobek*

for
Maggie & Esther Olivia

Preface

Scarcely a field in construction requires such close collaboration between all the parties involved in design, manufacture and execution as construction with textile materials. From the form finding and patterning to detailing and planning the erection, collaboration is essential between architect, engineer and the companies responsible for production and erection. This results above all from the particular mechanical behaviour of the materials used. Their composition and arrangement as load-bearing elements, the force relationships in the element and above all the capacity of such material to accept relatively high deformation all raise the question of the buildability of tensile surface structures – or membrane construction. This shows clearly how practicalities arising from the production of the materials and the erection of the structure exert a significant influence on the design process.

There has been continuous research and development for decades in the field of wide-span lightweight structures. The requirements placed on the design process by the needs of production and erection have in contrast only been investigated by isolated companies and consultants for application on particular projects. The present book is an attempt to remedy this lack of knowledge and, through the investigation of the current state of the art in the construction of tensile surface structures, represents an essential complement to the computer-aided calculation process of form finding and structural design.

The concentration on practical design details and their erection is used in this book to explain the basic interrelationship between manufacture and erection of flexible structural elements. Extensive specialist knowledge has been collected together and categorised in such a way as to be useful in practice. In some areas, optimisation potential has been pointed out for the process of designing membrane structures for practicality of erection.

Starting with the textile composition and internal structure of the flexible materials, the first part of the book explains their production processes, function and mode of operation as load-bearing and connecting elements. The mechanical behaviour of coated fabrics and the criteria for patterning are given special attention. Joint and connection methods for textile sheet elements and their special characteristics regarding manageability and buildability are also discussed.

In the second main part of the book, the importance of erection planning and its specific activities are explained with their economic and technological aims. This also includes measures to estimate and test erection activities and the scope of influence and purpose of erection planning in the individual phases of a project. The equipment used for erection is described, with an overview of the tools used for transport, lifting and tensioning processes. Tensioning systems for linear and sheet elements are described in detail.

Starting with production and construction technological parameters and their influential factors, procedures for the erection of some characteristic forms of mechanically pretensioned membrane structure are described and illustrated with examples from the practice. The description of the erection of flexible load-bearing elements includes essential activities like preparation, preassembly and erection on the construction site, and these are investigated and commented. The emphasis is on the process of introducing pretension into the membrane. Finally, there is an overview of methods for determining forces in flexible load-bearing elements.

The end result is a book, which will be of interest not only to architects and engineers concerned with the design and construction of membrane structures, but also students in relevant subjects, who should find the book useful as course material or for their own study.

It would scarcely have been possible to write this book without drawing on the experience of designers, manufacturers and builders. I would like to express my special thanks to all those who have supported me with suggestions and material in the past years.

I owe special thanks for personal interviews with Dr.-Ing. habil. Rainer Blum, Dipl.-Arch. Horst Dürr, DI Reiner Essrich and Univ. Prof. DI Dr. Karlheinz Wagner. I also wish to thank DI Peter Bauer, Dr.-Ing. E.h. Rudolf Bergermann, Ing. Christian Böhmer, DI Wilhelm Graf, Bruno Inauen, Stefan Lenk, DI Dr. Walter Siokola as well as DI Jürgen Trenkle.

Special thanks are also due for the expert support from O. Univ. Prof. i. R. DI Dr. Günter Ramberger in questions regarding wire rope manufacture.

I thank the following engineers for helpful correspondence: DI Dr. Herbert Fitz, DI Hansruedi Imgrüth, Ing. Wolfgang Rudorf-Witrin, DI Bernd Stimpfle, DI Rochus Teschner and DI Jörg Tritthardt.

Finally, personal thanks are due to those people and companies, who have supported and furthered the writing of this book by providing material.

These were Mag. arch. Silja Tillner, the engineers DI Christoph Ackermann, DI Benoit Fauchon, DI Knut Göppert, DI Hans Gropper, Udo Holtermann, DI Christian Jabornegg, Daniel Junker, DI Rudolf Kirth, DI Roland Mogk, DI Bruno Pirer, DI Michael Wiederspahn, DI Jürgen Winkler, DI Dr. Günther Zenkner, DI Dr. Rene Ziegler and also Tim Schubert and the following organisations: Arbeitskreis Textile Architektur, Bilfinger Berger AG, CANOBBIO SPA, CENO TEC GmbH, Covertex GmbH, Eichenhofer GmbH, Ferrari S.A., FIAB HF, Form TL Ingenieure für Tragwerk und Leichtbau GmbH, gmp – Architekten von Gerkan, Marg und Partner, Hightex GmbH, Histec Engineering AG, IF Ingenieurgemeinschaft Flächentragwerke, Inauen-Schätti AG, MERO-TSK International GmbH, Montageservice SL-GmbH, PFEIFER Seil- und Hebetechnik GmbH, Sattler AG, Schlaich Bergermann und Partner GmbH, Ingenieurbüro Teschner GmbH, Teufelberger GmbH, VSL Schweiz AG, Zeman & Co GmbH and Werner Sobek Ingenieure.

I also wish to thank at this point Em. O. Univ. Prof. Arch. Dipl.-Ing. Helmut Richter, the former head of the department Hochbau 2 at the Faculty of Architecture and Town Planning at the Vienna Technical University, whose confidence and stature as architect and teacher were an additional motivation for the creation of this work.

I wish to express heartfelt thanks to Ao. Univ. Prof. Dipl.-Arch. Dr. habil. Georg Suter for reading though the book and offering a critical discussion.

Vienna, August 2007 *Michael Seidel*

Contents

Foreword V
Preface VIII

1 Introduction 1
1.1 The importance of manufacture and erection 1
1.2 Ambition, aim and scheme of the book 2

2 Materials for tensile surface structures 5
2.1 Introduction 5
2.2 Structural composition and manufacture 6
2.2.1 Linear load-bearing elements 6
2.2.2 Surface load-bearing elements 26
2.3 Material behaviour of coated fabrics 39
2.3.1 Mechanical properties 39
2.4 Fabrication of coated fabrics 51
2.4.1 Development 51
2.4.2 Compensation, strip layout 53
2.4.3 Criteria for the patterning 53
2.4.4 Cutting out the pieces 63
2.5 Methods of jointing surfaces 65
2.5.1 Permanent surface joints 65
2.5.2 Reusable surface joints 71
2.6 Methods of transferring force at the edge 73
2.6.1 Geometry of the edging and effect on bearing behaviour 73
2.6.2 The detailing of edges and their anchorage at corners 74
2.6.3 Edge details 75
2.7 Corner details 81

3 Construction of tensile surface structures 85
3.1 Introduction 85
3.2 Construction management 87
3.2.1 Aims and tasks of construction management 87
3.2.2 Scheduling 88
3.2.3 Modelling erection procedures 91

3.2.4 Construction engineering 93
3.2.5 Design detailing for erection practicality 95
3.3 Erection equipment and machinery 97
3.3.1 Cranes and lifting devices 97
3.3.2 Tensioning devices and equipment for ropes 101
3.3.3 Tensioning devices and aids for membrane sheets ... 108
3.3.4 Scaffolding working platforms and temporary construction 112
3.4 Erection procedure 116
3.4.1 Criteria affecting the erection procedure 116
3.4.2 Remarks about the erection of the primary structure 127
3.4.3 Erection procedures for membrane structures 139
3.5 Construction 155
3.5.1 Preparation work and preassembly 156
3.5.2 Lifting and hanging structural elements 161
3.5.3 Introduction of loads – pretensioning 176
3.6 Control of the forces in flexible structural elements 193
3.6.1 Determination of force in ropes 193
3.6.2 Measurement of membrane stresses 195

4 Summary and outlook 197

References 199
Illustration acknowledgements 205
Projects (1989–2007) 209

1 Introduction

To analyse, formulate and finally substantiate a theoretical model requires complex processes and decision-making. Even though such interacting processes with the intention to *make* or *manufacture* something are hard to illustrate with diagrams, solutions still have to be looked for and developed and the processes notionally fixed, illustrated and explained in advance.[1] In the construction industry, where materials are arranged to form structures, these processes are usually called conception, design and construction.

The term *construction* here means the route developed to materialise such theoretical models through the conception, design and construction process. The final structure therefore represents a considered and materialised entity. This book understands the construction process as the *material arrangement of the environment* and the materials required are found the appropriate space for consideration.

To reduce the rules of construction to the overcoming of materials and methods alone does not achieve the required purpose. The essential influence of manufacturing and erection methods in the determination of the most suitable type of construction or structure should not be underestimated. It makes more sense to regard the technology of construction as a fundamental source of income, which becomes richer with repeated use.[2]

1.1 The importance of manufacture and erection

The erection of a structure requires a wide-ranging consideration of methods, procedures and execution. The aim of these considerations is the creation of a qualitatively high-value structure and the achievement of short erection times. This can be achieved through a high degree of pre-fabrication and the development of efficient erection procedures.

To be able to reach the right decisions about the optimisation of time, cost and energy at each of the various project phases requires extensive technical knowledge and experience. The effective collaboration of all the specialists involved in the design and implementation is also vital. They are all significantly involved in the implementation of the physical design.

As a result of progress in the development of materials and components in construction, the understanding of the methods for producing and jointing materials and their technical specification is becoming ever more important for the engineers and architects involved in design. Manufacture and erection on site should therefore be regarded as core processes in the development of structural systems.

The respect for considerations of manufacturing and assembly has a long tradition in the design of construction components. Particularly in steelwork, timber and precast concrete construction, the layout of construction elements is influenced not only by architectural formulation and structural demands, but also by factors derived from the practicalities of production and assembly. The elements are prepared at the works, partially pre-assembled, numbered and transported to the construction site, where they are put together to form units for erection in a further preassembly process.

In the field of membrane structures, a specialised area of structural engineering where textile structures of fabric and ropes form wide-span roofs, this is also the usual practice on site. One reason for this is the stringent requirements on the manufacture and testing of the materials, which makes production on site impossible and demands particular erection technologies.

Membrane structures, on account of their external form and their light weight, are categorised under wide-span lightweight structures. Regarding their structural behaviour, they differ from conventional structures above all in that the external forces are transferred exclusively through tension. They are therefore described as form-active systems. With the appropriate choice of materials, curvature and force transfer, their structural form corresponds exactly to the line of the forces.[3]

Coated fabrics are usually used for the textile surface elements of mechanically tensioned membrane structures. To improve their material properties, they are made of composites of various materials. The wire ropes used as linear struc-

---

1 *Ferguson, E. S. (1993)*
2 *Buckminster Fuller, R. (1973)*
3 *Engel, H. (1997)*

tural elements and their connections are for functional reasons also composed of various components.

Surface and edge elements under tension have negligible bending strength; they are assumed to be flexible materials. Their geometrical arrangement and formation as bearing elements in a membrane structure, which permit relatively high deformations under the forces acting on the element and the capacity of the material, demand answers to the question of the "buildability" of such a structural design.

In addition to the local conditions on the construction site and the choice of suitable equipment, it is above all the rules of construction methodology for the erection of structural elements under tension, which require understanding of the internal composition and mechanical behaviour of the material to be used and enable the implementation of the chosen assembly method.

In contrast to conventional building methods, where assembly consists of the addition of further elements horizontally or vertically, a membrane structure only achieves sufficient stiffness in its structural elements and in the structural system after the application of pretension to the linear and surface elements.

The erection process to be used for a membrane structure is therefore decisively dependant on the production quality of the materials used. The original shape, the deflected shape and above all the jointing process have a strong interaction with the method of erection and offer important potential for optimising the way of constructing such a structure.

1.2 Ambition, aim and scheme of the book

The thousand-year history of building with membrane surfaces was restarted in the second half of the 20th century. In Europe, it was mainly the engineers, architects and scientists in the circle around the German architect Frei Otto, who between 1970 and 1985, as part of the widespread scientific cooperation in and outside Germany under the auspices of special research area 64 in Stuttgart, achieved a major contribution to the research and development of wide-span membrane structures. The companies who supported these developments should also be mentioned.

Since then, there has been further continuous development in membrane construction. The use of computer-aided calculation methods to solve problems of form finding and statics for thin-walled surfaces and also the development of new materials and methods of connection, taking basic parameters into account, today enable a mostly systematic design of membrane structures. If, however, one is looking for answers to questions about the practical implementation of membrane structures today, these are mostly only to be found in reports of completed projects, in which the site erection process is only briefly described. A few exceptional examples have been presented to the relevant specialists.

The purpose of the present book is therefore the systematic investigation of the current state of the technology for implementing the construction of wide-span lightweight surface structures.

The orientation on actual construction and its erection should allow basic interactions to be recognised; how the processes of production, delivery and jointing all influence one another and offer potential for the optimisation of the design process. Of special interest is the explanation of the interaction between the production of the individual materials.

Instruction in the methodology of the manufacture and erection of membrane structures is not part of this book. It also represents no simple compilation of descriptions of the technology. The emphasis is more the presentation of the complex areas of influence, dependencies and connections between the design and construction of membrane structures. Analytic questions of form finding, the economic aspects of resource planning, capacity management and estimating can only be described to a limited extent.

The book is divided into two main parts. The first of these describes the materials used to form structural elements and their production processes. The selection criterion is a categorisation of construction elements according to geometry and function. Starting from their material and structural composition, the materials used are described and their production processes are explained. Examples are given of their function as structural elements and the principal correlations between the requirements of geometry and bearing behaviour, and also details of their composition.

To make clear the complex material behaviour of coated fabrics, the basic mechanical properties of the most commonly used types of fabric are summarised and their influence on practical implementation in construction is explained.

Then possible details for surface joints, edges and connections to other components are discussed and illustrated with examples.

The second main part of the book deals with the erection of membrane structures and their load-bearing elements. Starting with the scheduling of the construction progress, the most important construction equipment is discussed. The emphasis here is on devices and equipment for tensioning wire ropes and membranes. Starting from the parameters influencing the principle of erection, methods of erecting membrane structures are described and erection operations investigated systematically. The characteristic stages of the individual processes are described and illustrated with diagrams. This is supplemented with photographs of erection work on completed projects.

Finally, the process of assembly and erection and the procedures on the construction site are explained. All the essential steps in the erection of structures are explained, from the preparation work and the pre-assembly through lifting, hanging and tensioning. The emphasis here is the process of introducing the forces into the membrane surface. In the last section, there is a summary of the methods and procedures for measuring forces in wire ropes and membranes.

The closing discussion comments on open questions and offers a view of possible developments in the future.

2 Materials for tensile surface structures

2.1 Introduction

The chief characteristic of surface structures is the large clear spans, which can be roofed over very economically without internal support. Construction forms, which transfer all external forces as tension, have proved most successful in this specialised area of construction, with the exception of shell construction. Form-active tension systems have the advantage compared with traditional structures that, given corresponding interaction of form, force and material, the volume used is minimised without reducing the strength, stiffness or stability of a structure.

Membrane surfaces installed in form-active structures are composed of load-bearing elements of varying materials, types of construction and geometry. Normally these are large jointed surface elements and linear edge elements. Both types of element can only bear loading in a particular shape (curved), and have to fulfil certain technical criteria. In addition to keeping the weather out, they have to be resistant against chemical and biological attack and must be non-flammable in case of fire. Concerning structural safety, they have to conform to specific weight and balance requirements and above all strength and stiffness requirements, and also enable the stress distribution to be matched to external loading conditions. Problems of force transfer have to be dealt with correctly considering the stiffness behaviour. The construction material of the load-bearing elements has to be formed and dimensioned to suit their purpose in accordance with these requirements.

The dimensions of the surface elements are exceptionally two-dimensional. Linear load-bearing elements are used to transfer loading though the edge. The mechanical material properties of both types of load-bearing element have to enable the load transfer from the multiply curved surface forms exclusively through high tension forces. The types of material used normally have high elongation stiffness and negligible bending stiffness. The mechanical behaviour of the materials in buckling and plate buckling is, depending on the loading, relatively *flexible* in comparison with the structural system.

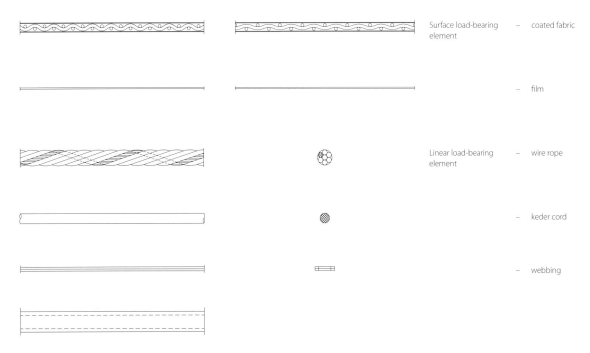

Fig. 1: Flexible load-bearing elements in wide-span surface structures

Two groups of material are mostly available today as membranes for planar load-bearing elements (Fig. 1). Either composite textile materials as coated or uncoated fabrics with synthetically manufactured and processed fibres, which are described as technical textiles; or fluorocarbon polymers as extruded films, which are described as technical plastics. The linear bearing elements can be tension members of steel wire rope, textile webbing or extruded keder cords. The above-named materials are normally processed into working material using industrial production processes like casting, forming and joining.

The behaviour of a material when used as a load-bearing element plays an essential role in the successful implementation of a wide-span surface structure. The important properties for the mechanical behaviour of a load-bearing element are the dimensions, material properties and composition. This leads to the conclusion that knowledge of the manufacturing process and material properties of the material to be used is a precondition for the production of a design with convincing layout and functionality.

The layout of the contents of the current chapter
This chapter is laid out according to the material of a flexible load-bearing element. The categories are divided according to the geometry and function of load-bearing building components. This division is intended to explain the complex interactions between production, joining and their influencing parameters on the mechanical behaviour of the material.

Starting with the composition and internal structure, the materials mostly used for the construction of wide-span surface structures are presented and their method of production explained. Details are given of the various states in the various manufacturing processes as well as the reciprocal effects of the mechanical properties between manufacture and erection.

To explain the complex material behaviour of the flexible surface element, the mechanical behaviour relevant for the characterisation of deformation of the commonly used coated fabrics is explained systematically under the effects of loading, time and temperature.

One design stage of the greatest importance for the practical implementation on site is the patterning, that is the calculation of the required shapes for pre-cutting. The constraints on the mechanics of the flexible surface element resulting from the shape, the load-bearing behaviour and the assembly are closely related to and are reciprocally affected by the type of pattern. Additionally, the topological, static, production and assembly criteria for the patterning of flexible planar elements are described and investigated for their buildability.

The various methods of transferring the forces into the membrane material require corresponding detailing of the edges and corners as well as special methods of fixing to the rigid construction elements in order to stabilise the membrane surface in the required form. The most important parameters and design principles for the geometry and type of the edging are also discussed, as is their effect on the load-bearing behaviour of the membrane surface. In addition to this, connection details to neighbouring elements at the edge and corners of partial surfaces and jointing techniques at assembly joints are described and illustrated with examples.

2.2 Structural composition and manufacture

2.2.1 Linear load-bearing elements

Membrane surfaces have to be stabilised in their location by closed edging. The tension forces acting in the plane of the membrane are transferred into adjacent elements, where they can be carried to the foundation. Flexible linear tension members can be installed as an edge detail to the surface, either to reinforce the edge or to be solely responsible for conducting the edge loads tangential to the edge curve of the membrane surface. This is then called a flexible edge. The axial dimension of such an inserted tension member is many times larger than its other dimensions. They are only stiff in the axial direction and are loaded uniformly over their cross-section and exclusively in tension. For wide-span structures, they are curved in two dimensions, or in special cases in three dimensions. With the appropriate geometry and detailing, deformation in the plane of the membrane can be partially resisted by the deformation of these flexible edge elements. These edge elements normally consist of spiral steel wire ropes and band-shaped textile webbing belts.

To reinforce the edge of a textile surface, this can also be constructed rigid. This is called a rigid edge. Cord-shaped plastic keders are used in such edge details for load transfer to the rigid metal fittings.

The following section gives an overview of the composition and the types of linear structural elements, which are used in wide-span planar structures, with a description of the industrial processes for their manufacture. Construction details for anchoring to adjacent construction components are also discussed.

2.2.1.1 Wire ropes

The wire rope is a machine element subject to a range of loadings.[1] It consists of aligned and stretched wires and can support tension forces constantly or dynamically.[2]

The highly flexible tension members used in materials handling, which run over rollers, sheaves or drums, are described as *running ropes*. Running ropes would mostly only be used for convertible wide-span surface structures; these are not described in detail here. Cables or bundled wire ropes, as are used in bridge building, are not considered here either.

Tension members made of steel wires, which are used in construction to support static forces, are called ***static ropes.*** They can be use in two-dimensional and in three-dimensional tension systems. Wires with tension strength of up to about 1,770 N/mm^2 are used for standing ropes.[3] This is achieved through thermal and mechanical treatment during production. In addition to the high strengths available, the use of wire ropes has major advantages above all during erection. Efficient construction is possible under practically all climatic conditions.

In structures under tension loading, where at least a part of the loading is supported through the deflection of tension members, static wire ropes and bundles of wire ropes fulfil important roles as single structural elements and as part of systems. As tension elements for edges, stay ropes and supporting ropes, they must have the required construction and mechanical properties to fulfil the specification. Rope wires already undergo considerable loading during manufacture, a combination of tension, bending, torsion and compression. In the structure, static ropes have to show high strain stiffness for many decades, support any forces arising from deflection or shear compression and be sufficiently protected against corrosion. These requirements have led in the course of ongoing technical development to various constructional details, which are summarised and described below.

The already mentioned special research project 64 "Widespan Lightweight Structures" made an essential contribution to research into the use of wire ropes in building structures in various project areas. The results from the special research project have been used repeatedly as the basis for German standards and guidelines.

Construction of ropes

Ropes consist essentially of wires, cores and strands.

The insert (core) lies in the centre of the strands and ropes and serves to bed and support the wires, strands or ropes. The two types are fibre inserts and steel inserts. Static ropes mostly have steel inserts, which ensure stiff bonding of the wires.[4]

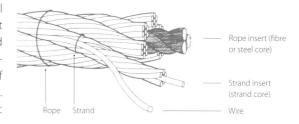

Fig. 2: Construction of a rope

The steel wire, the smallest component of the rope (dia. 0.5 – 7.0 mm) is mostly made of unalloyed carbon steel. The starting material for rope wires is rolled wire, which can be hardened by various shaping processes. These processes are cold drawing, die drawing or roll drawing. Controlled heat treatment before or between the individual drawing processes and subsequent quenching (patenting) produce higher strengths. The combination of both manufacturing processes and the intended surface quality considerably improves the quality of the rope wire.

Strands are spun helically around an insert (strand core) and consist of one or more layers of wires. Strand cores consist either of one wire (core wire) or of spun yarn.

The construction of the strands has a great influence on the properties of a rope. Strands can be categorised according

round waisted wedged Z profile triangular flat oval

Fig. 3: Wire sections

1 *Stauske, D. (1990)*
2 *Schefer, M. (1994)*
3 *Stauske, D. (2000)*
4 *Gabriel, K. (1990)*

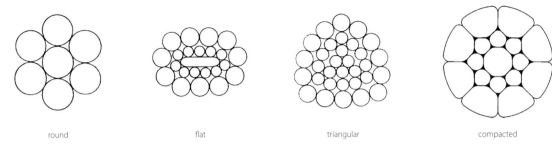

round   flat   triangular   compacted

Fig. 4: Types of strand

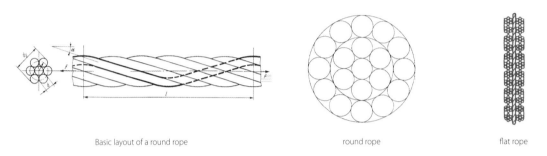

Basic layout of a round rope   round rope   flat rope

Fig. 5: Types of rope

to their shape into round strands, shaped strands and compacted strands.

Ropes consist of one or more layers of wires or strands, which are spun helically round a core. They can be categorised according to their shape into round ropes and flat ropes.[1]

Single parallel members of wires or strands, combined to form larger units, are called *bundles.* These can be categorised into parallel wire bundles and parallel strand bundles (strand bundles). Bundles are mainly used as suspension cables of bridges. On the construction site, parallel wire bundles are bound at intervals with soft iron round wire or clamped with clamps to hold the bundle together.

Parallel single members as ropes or bundles, combined to form larger units, can also be called *cables* (rope bundles). Cables are mostly used in building for larger spans, for example the perimeter rope bundle of a spoked wheel roof construction.

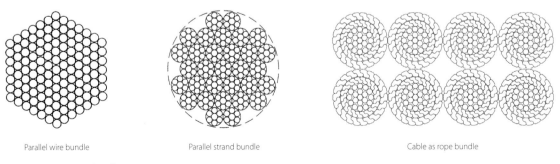

Parallel wire bundle   Parallel strand bundle   Cable as rope bundle

Fig. 6: Wire, strand and rope bundle

---

1   Definitions according to the Austrian standard ÖNORM M 9500

Lay type and direction

The method of rope making is also called spinning. Ropes are categorised according to the lay direction of the strands and the lay direction of the rope.

The *lay direction* of the strands (z or s) is the direction of the helical line of the outer wires with reference to the axial axis of the strand. The lay direction of the rope (Z or S) is the direction of the helical line of the strands, the outer wires in a spiral rope or the component parts of a cable-laid rope, related to the axial axis of the rope.[1] The terms s, S, z and Z are illustrated in Fig. 7.

The *type of lay* describes the lay direction of the wires in the strand and of the strands in the rope. Stranded ropes can be spun with regular lay (sZ or zS), where the wires in the strands have the opposite lay direction to the strands in the rope, or Lang's lay (zZ or sS), with the strand wires having the same lay direction as the strands in the rope.

Ropes with Lang's lay are exceptionally hardwearing and more flexible than regular laid ropes. Ropes with regular lay have less tendency to twist and spring, are less susceptible to damage by dirt and deformation and less flexible than ropes with Lang's lay.

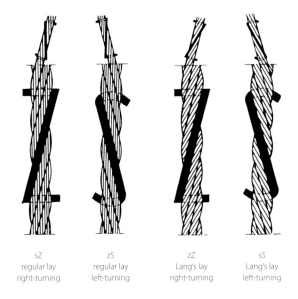

| sZ | zS | zZ | sS |
| regular lay right-turning | regular lay left-turning | Lang's lay right-turning | Lang's lay left-turning |

Fig. 7: Lay type and direction

Lay length and lay angle

The *Lay length* describes the length of wire in one rotation about the rope axis. There is the following relationship between the lay l_L and the lay angle α_L:

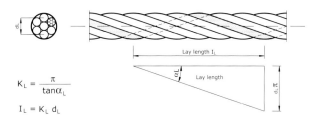

$$K_L = \frac{\pi}{\tan\alpha_L}$$

$$I_L = K_L \, d_L$$

Fig. 8: Lay angle and lay

The lay number K_L is the quotient of the lay length and the diameter measured to the strand centres.

Types of rope

The tension members of cold-drawn high-strength steel, which are used in the construction of lightweight surface structures, are divided into spiral ropes and stranded ropes. The following types are mostly used:

Open spiral strands (OSS) consist of spirally spun round wires with almost identical diameters, which are spun round a core wire in many layers, usually spun in alternate directions. They have a medium density (proportion of steel area to the overall cross-section) and, dependant on wire diameter, a more or less uneven surface. Open spiral strand ropes are produced with up to 91 single wires. With a higher number of wires, the geometrical strength of the rope bonding is considerable reduced, and fully locked ropes are better. Open spiral strands are suitable for light to medium forces and are frequently used for edge ropes in membrane construction and as supporting ropes or in rope truss construction.

Half-locked spiral strands (HVS) have a half-locked layer of round and waisted wires. They were a precursor of the fully locked ropes and are seldom used for static ropes nowadays.

Fully locked spiral strands (VVS) consist of a core of round wires and one or more layers of shaped wires. The mostly Z-shaped form of the external wires results in a dense locked surface, resulting in good preconditions for corrosion protection measures. They have an extremely flat external contour to protect the inside of the rope from water ingress or aggressive media and also prevent leaking of the rope filling. The external arrangement of shaped wires also improves the mechanical protection. The higher metal cross-section permits the support of higher loads with low dead weight.

Because of their construction, they can be manufactured practically non-rotating. For permanent structures, they are used mainly as carrier and tensioning ropes for lightweight structures or as stay ropes for masts.[2]

---

1 Definition according to European Standard EN 12385-2

2 Westerhoff, D. (1989)

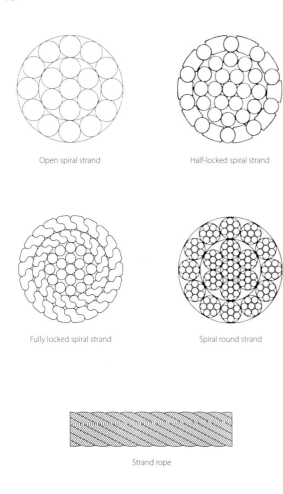

Fig. 9: Types of wire rope

Round spiral strand ropes are formed of many strands, which are spun spirally in regular or Lang's lay in one or more layers around an insert. Strands running next to each other can have opposing lay directions. Round spiral ropes are mostly used when more flexible ropes are required. They have, however, a relatively low density and a fissured surface, which makes them easy to handle, but they are more susceptible to corrosion and wear than spiral ropes.

Round spiral strand ropes are often used on account of their easy handling as bracing, stay wires and low-level truss tension members for temporary buildings, or as handrail ropes in stairs, balconies and bridge parapets.

Rope inserts: For static wire ropes, steel inserts are normally used. Running ropes normally have fibre inserts arranged between the individual layers, in order to prevent the strands sliding against each other.[1]

Manufacture of static ropes and bundles

Metal tension members must have high strain stiffness, must not come apart at the defection location and have to be protected against corrosion. They must therefore be manufactured with as high a packing density as possible.

The manufacture of wire ropes takes place in stages. The individual processes range from unwinding the wire through stranding the wires and spinning the strands to the preparation of the final product for delivery. This requires great care, production controls and extensive quality assurance measures art every stage of production.

Spinning process and type

Ropes are made by twisting wires in layers around a core wire. This process is called spinning. The process of *spinning* is done with a strander, where the wires are formed to a spiral. According to the type of rope, rope cross-section and required production speed, this can be done using prefabricat-

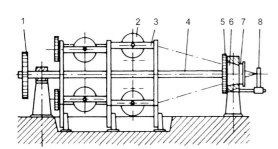

Fig. 10: Stranding basket of a planetary stranding machine

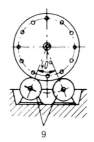

1 main drive
2 bobbin
3 stranding spider
4 hollow spindle
5 drive for pull-out
6 stranding head
7 separating plate
8 closing die
9 support rollers

1 *Gabriel, K. (1990)*

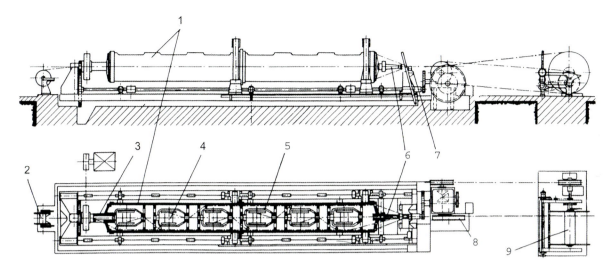

1 stranding tube	4 bobbin	7 closing die
2 inlet bobbin	5 bobbin flyer	8 transfer sheave
3 drive shaft	6 output arm	9 winder

Fig. 11: Tubular wire rope stranding machine

ed strands on a planetary stranding machine (Fig. 10), tubular stranding machine or flyer stranding machine. Each layer of wire is spun in its own work stage. To almost completely relieve the twisting caused by the spinning process, the wire layers are spun alternatively left and right-handed.

Ropes with many layers and larger diameters are mostly spun using a planetary stranding machine running at a low revolution speed (about 50 rpm). The steel wire rolled on the bobbins is fixed to the stranding spider rotating around the shaft. The rope insert runs through a hollow spindle to the closing die, from where the wires coming from the spider are wound around the insert. The advantage of rope making with a planetary strander is mainly that larger wire bobbins can be used.

Ropes of smaller diameter can be spun using tubular stranding machines, also called quick stranders, running at up to about 3,000 rpm (Fig. 11).

The wire bobbins in this machine are arranged in a row along the axial axis of a rotating tube. The bearings of the bobbins are arranged so that they do not turn with the tube.

The wire is fed towards the rotational axis of the tube, where it is deflected to be fed along the tube to the closing die.

The wires are brought together in both types of machine at the closing die, where they are fixed in position by the twisting of the individual strands in a roller system. The strands are deformed plastically, which is described as ***preforming.*** Running through centring rollers behind the strander head is called ***post-forming.*** After post-forming, the rope has its final form and, after passing over a transfer sheave, is wound onto a rope drum. These two alignment processes are intended to relieve the internal stresses caused by the stranding process and to produce stress-free or low-stress ropes.[1] Winding the wire onto the bobbins, unwinding, deflecting, aligning

Fig. 12: *left:* transfer sheave; *middle:* planetary stranding machine; *right:* model of a stranding mechanism

1 Trurnit, P. D. (1981)

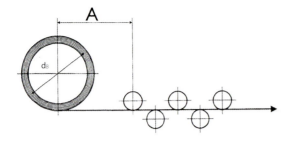

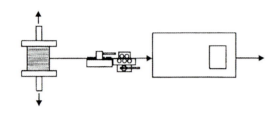

Fig. 13: *above:* Schematic diagram of alignment rollers; *below:* Schematic diagram of reel alignment mechanism

and stranding all have a considerable influence on the final quality of the rope.

The single wires are processed to strands in the stranding process. The single strands can then be further stranded together in following steps. According to the number of successive stranding processes, this produces once, twice, three times of four times stranded rope. Strands, which have been stranded once, and spiral ropes can be processed in further steps to make stranded ropes or spiral strand ropes. A number of strands are spun to form a rope core, around which, in the following final rope-making process, the outer strands are spun. These can be either parallel or cross-laid (standard) according to lay and lay angle of the individual layers to each other.

In cross-laid ropes (a in Fig. 14), all wires have the same direction of lay and the same lay angle, with the wire layers having differing lay lengths. The stranding is done with round wires of the same diameter. The crossing of the wires produces load transfer at points between the adjacent layers of wire (c in Fig. 14). In parallel laid strands (b in Fig. 14), all wires have the same length of lay and lay direction. The lay angle, wire diameter and wire lengths are all different. The load transfer is linear (d in Fig. 14).

Further types of stranding are extended parallel lay, composite strands, and contra-laid.[1]

The helical arrangement of the wires in a rope cause additional stresses, strains and compression. The difference between the calculated breaking load and the actual breaking load is described as the *stranding loss.* The *stranding factor* is an empirical value for the calculation of the real breaking load, which takes the stranding loss into account and is quoted by the manufacturer.

High breaking loads for ropes can only be achieved when the strength of the wire material, the stranding factor and the metal cross-section of the rope are as high as possible. The ratio of the metal cross-section of a rope to the cross-section of the circle around it is given as the *filling factor.*

Twisting and bending behaviour of ropes

Traditional wire ropes try to untwist under load. The tendency of a rope to twist is composed of two components, the manufacturing twist and the loading twist. On the one hand, re-

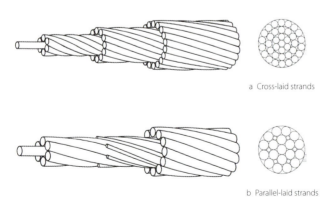

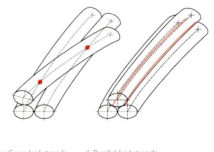

Fig. 14: Lay types according to Thyssen

1 Westerhof, D. (1989)

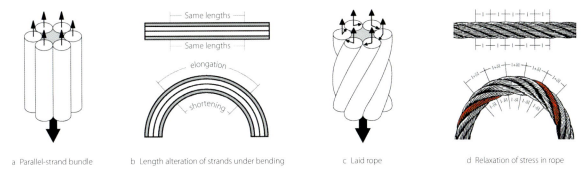

Fig. 15: Length alteration and stress relaxation in deflected ropes

sidual elastic stresses remain in the rope as a result of plastic deformation. On the other hand, loading of the rope, causing axial deformation in the direction of the wire axis, leads to a reduction of the angle of lay, which causes a torque about the central axis.[1]

In order to avoid the tendency of a rope to twist under loading, the torque can be compensated and almost eliminated by spinning the individual layers of wire in alternate directions. The direction, in which the rope is laid, leads to the descriptions *left-hand lay* or *right-hand lay*.

According to the load-dependant twisting behaviour or whether the rope applies a torque to its fixings, a rope can be described as *rotation-resistant* or *rotation-free*. A rotation-free rope has a core rope, which is spun in the opposing direction to the outer strands. A rope is described as rotation-resistant …" if the free end of the rope under load turns only slightly or when the fixed end of the rope only applies a slight torque to its fixing".[2] A rope is described as rotation-free … " if the unsupported rope end does not turn under loading or the fixed end applies practically no torque to its fixing".[3]

When bundles of parallel strands are deflected (bent), the outer strands are stretched more and the inner strands are compressed, that is partially or fully relieved of load. In the case of a rope, which is spirally wound around a core, each strand is only stretched or compressed in places. (d in Fig. 15).

In this way, a large proportion of these extensions and forces caused by bending can be relieved. The bending behaviour of wound ropes is considerably better than those consisting of bundles of parallel strands.[4] Parallel wire and strand bundles can be assembled in the works or on the construction site.

The essential parameters for the bending behaviour of rope tension members are the defection length, the mechanical lock, the wire diameter, the lay length and the amount of shear compression.[5]

Compaction

The bearing capacity of a rope increases with the cross-sectional area and the strength. If thin external diameters and large metal cross-section are required, a compacted rope can be used. Compacted ropes can reach filling factors of up to 95 %. In order to reduce the air voids between the wires, ropes can be axially drawn or rolled after the stranding process and additionally compressed from the side and deformed. Ropes subjected to such compaction methods are called *compacted* ropes.[6]

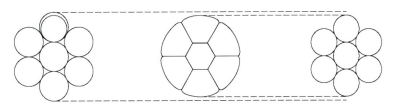

Round wire strands before compaction   Compacted strands   Round wire strands with the same Ø

Fig. 16: Compacted strands

1   Schefer, M. (1994)
2   Singenstroth, F. (1998)
3   Singenstroth, F. (1998)
4   Verreet, R. (1996-1)
5   Gabriel, K.; Wagner, R. (1992)
6   Ramberger, G. (2003)

Compacted ropes are used mainly as running ropes. When they run over pulleys, the advantages of increased seating area and lower friction in the pulley groove become clear. Further advantages are increased breaking load, the continuous bearing stress and less friction between the inner and outer layers of the rope.

Pre-stretching

When a rope is loaded, this causes permanent strain. In order to avoid these strains in advance, increase the deformation modulus of the rope and determine the exact length of the rope under load, ropes are pre-stretched with the working load or more in a pre-stretching plant. In pre-stretching tracks up to 250 m long, ropes can be tensioned up to a load of up to about 6,000 kN. The required tension force is transferred to the rope with clamps. Longer ropes are pre-stretched in sections. In order to achieve a uniform strain behaviour and thus sufficient strain stiffness under increased tension, ropes are repeatedly loaded in a pre-stretching plant (dynamic pre-stretching).[1]

The tension force applied during pre-stretching brings the wires in the rope into their optimal arrangement for stressing. At isolated highly stressed locations, the plastic limit of the material is exceeded. This procedure helps to reduce local peaks of stress and to compensate the wire lengths.[2]

Cutting ropes to length

After the rope has been made, it must be fabricated. To do this, it must be shortened in stress-free condition by the extension caused by pre-stretching. In addition to the rope extension resulting from pre-stretching, shortening due to potential deflections of mast and foundations have to be taken into account. The required compensation data is delivered by the engineer in no-deformation cutting drawings. Because the length of a stress-free rope cannot be measured, the rope is loaded to an agreed value after pre-stretching and dynamic pre-stretching, and then measured, marked and cut to length.

The exact seating of the end connection must also be marked under load, and also an axial line, which enables the twist-free assembly of the end connection and the rope.[3] Predictable (plastic) creep deformations are also taken into consideration before cutting to length.

Corrosion protection of ropes

High-strength ropes react very sensitively to mechanical damage (notching) and material loss (wear, corrosion). Damage is repairable only with difficulty or not at all, so tension members should be so well protected that corrosion cannot start. The design of structures should also make it possible to exchange tension members.[4]

Corrosion protection measures for wire ropes can essentially be divided into four categories. The individual wires can be protected, the rope internally or the rope externally. The corrosion protection can also be achieved or improved by appropriate design detailing.

Corrosion protection of individual wires

The usual corrosion protection for rope wires is galvanising. Some manufacturers offer ropes with wires protected by Galfan, a eutectic alloy of zinc (95 %) and aluminium (5 %). On account of the considerably better course of corrosion with Galfan coating, Galfan galvanising can offer longer resistance than normal galvanising under the same conditions. If the outer layer of the protection is damaged, the corrosion protection effect can regenerate again from the lower layers. The special pliability and good adhesion of the Galfan coating also makes ropes easier to manufacture.[5] Round wires up to about 3 mm can be made from non-rusting stainless steel wires.

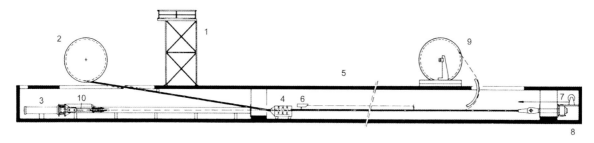

Fig. 17: Diagram of a pre-stretching plant

1 Tower for filling compound
2 Reel, unwinding trestle
3 Tensioning cylinder
4 Clamping carriage
5 Stretching equipment
6 Length measurement equipment
7 Capstan
8 Rope thrust bearing
9 Reel, winding trestle
10 Measurement station for pretension

---

1 Stauske, D. (2000)
2 Oplatka, G. (1983)
3 Stauske, D. (2000)
4 Klopfer, H. (1981)
5 Stauske, D. (1995)

Fig. 18: Pouring the lubricant

Corrosion protection of the rope internally

Wire ropes made of galvanised wires will loose their zinc layer in a foreseeable time without further corrosion protection measures. This can be prevented by the application of a suitable filling compound (paste) in the voids between the wires and the strands, which ensures the corrosion protection of the rope internally. This process is known as filling or lubricating. Filling reduces the internal friction between the individual wires and strands. The most commonly used lubricant is zinc dust dissolved in synthetic oil. The lubricant is added immediately before the closing die and dosed to hinder it being exuded later under pretension (Fig. 18). When stranding stainless steel, a thin film of linseed oil is normally used to prevent overheating of the closing die.

Corrosion protection of the rope externally

If serious damage of the corrosion protection is to be expected during erection of the ropes, then they can be painted for additional corrosion protection after erection. This is typically 4–6 layers of pigmented or unpigmented plastic paint, applied after the rope has been loaded. According with the manufacturer's instructions, it is not necessary to apply additional corrosion protection measures to ropes with Galfan coating.[1]

Corrosion protection through design detailing

Particular emphasis should be placed on correct detailing to ensure the protection of connections from corrosion. Poorly accessible voids and gaps around clamps, saddles, end anchorages and loops etc. should be sealed with permanently elastic material.[2] In principle, care should be taken that rainwater running along the rope from the anchor bush to the entrance of the rope into the end anchorage can be so diverted that it can permanently drain freely and will not cause corrosion.

When using ropes of high-alloy rustproof steels, the environmental conditions should be taken into consideration. If the protection layer is damaged, the combination of moisture and oxygen can result in corrosion danger.

Rope anchorages

The forces, which act in tensioned wire ropes, must be transferred further along the free length of the rope or resisted by an anchorage. The force transfer can be through friction, formfit or bonded via intermediate anchorages and end anchorages to other construction elements, typically masts, columns, edge beams or foundations. The detailing of the anchorage depends on the type and diameter of the rope, the extent of the forces to be connected, the type of connection, the erection process and the requirements of future maintenance.

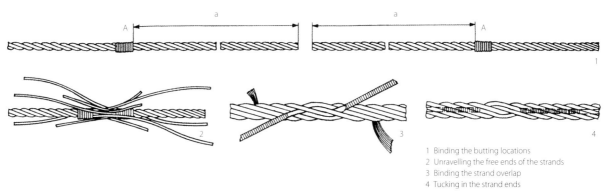

1 Binding the butting locations
2 Unravelling the free ends of the strands
3 Binding the strand overlap
4 Tucking in the strand ends

Fig. 19: Making a longitudinal splice

---

1   PFEIFER company material (2003)
2   Peil, U. (2002)

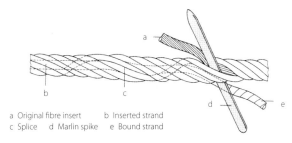 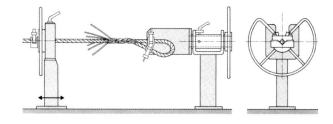

a Original fibre insert   b Inserted strand
c Splice   d Marlin spike   e Bound strand

Fig. 20: *left:* Splice; *right:* Splicing frame

End-to-end connections

The oldest form of end-to-end connection for ropes is the splice. Splices can connect ropes with themselves or to others.

To make a splice, the free ends of the rope are unravelled and braided with the end of the other rope. The strands of the rope ends are exchanged alternatively, with an unravelled strand being replaced by the corresponding strand from the rope to be connected. The free rope ends are bound with thin wire (A in Fig. 19) and worked into the rope (1–4 in Fig. 19). Splices for intermediate and end-to end connections are only possible with stranded ropes.

Splicing is normally done without special machines. For steel ropes, the main tool is a flat marlinspike, which is pushed into the rope to make an opening for the strand to be inserted. Heavy ropes are hard to open. For this purpose, splicing frames have been developed, in which the rope can be fixed (right in Fig. 20). The rope can be opened by turning the clamped ends against each other.[1] The splicing of ropes is a lot of work and requires a special training.

End-to-end splices are used above all where endless ropes are needed. This is mostly the case with running ropes. The advantage of this type of connection is that the splice leads to no thickening of the rope and can be run over rollers and pulleys. Longer ropes with local damage can be repaired by splicing.[2]

End-to-end connections of static ropes can be made for temporary purposes with bolted multi-part wire rope clips. The force, which can be transferred, depends on the sliding force resisting the clip from sliding on the rope.[3] When the nuts are tightened, the saddle presses the two rope strands against each other, and the rope ends are connected by friction and also by mechanical interlocking.

The number of clips required is specified at 3–8, with 2–5 further clips being fixed at a spacing according to rope diameter of at least one clip width. For 7 mm diameter rope, 3 clamps are normally used, and 8 clips for 28 mm diameter ropes. If the rope diameter is above 40 mm, rope clips are not suitable.[4]

Before assembly, the threads of the U-bolt and the bearing surfaces of the nuts are greased to ensure frictionless tightening. After tightening by hand, the clips are tightened with a torque wrench. Because the rope diameter reduces under load, the clips must be tightened again after applying the load[5].

The connection of wire ropes bearing high forces in their axial axis can be made with speltered sockets or with connectors like links, threaded rod or wedged sockets.

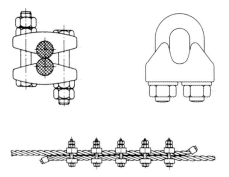

Fig. 21: *above left:* Heuer-Hammer clamping system;
above right: Rope clip according to DIN 1142;
below: End-to-end connection with rope clips

---

1 *Feyrer, K. (1986)*
2 *Scheffler, M. (1994)*
3 *Stauske, D. (2002)*
4 *Peil, U. (2002)*
5 *Verreet, R. (1996-1)*

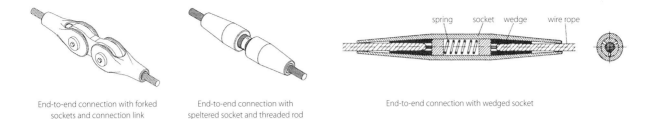

Fig. 22: High-strength end-to-end connections

End connections

To transmit forces to other construction elements, forces must be transferred into and out of the relatively thin wires of the rope with anchorages as short as possible. These rope end connections can be categorised into those which hold due to friction and material bond, and those, which hold through friction and form-fit (Fig. 23).

The end connections for static high-strength rope tension members for wide-span lightweight structures are mostly permanently speltered sockets or swaged connections. Spliced and wedged connections are only seldom used, and so are not handled further here.

Rope sockets

Open spiral wire ropes over 36 mm diameter and fully locked ropes are normally anchored in conical sockets by speltering. The sockets consist of thick-walled, high-strength wrought iron or cast steel with conical inner walls. The force transmission from wire rope to the cone of the casting is through

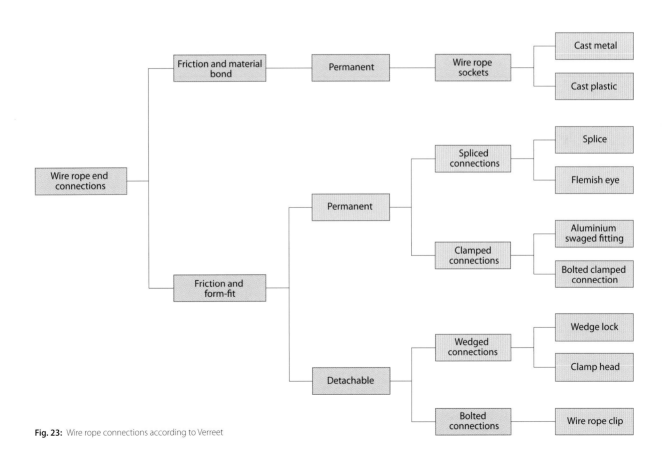

Fig. 23: Wire rope connections according to Verreet

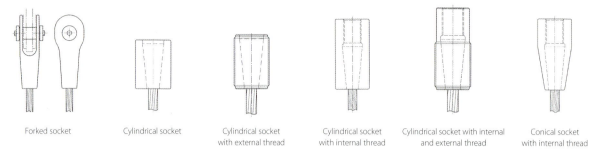

| Forked socket | Cylindrical socket | Cylindrical socket with external thread | Cylindrical socket with internal thread | Cylindrical socket with internal and external thread | Conical socket with internal thread |

Fig. 24: Types of speltered sockets

friction and chemical-mechanical material bond, and from the casting to the body of the socket is through form-fit. The poured filling of the socket may be either metallic or plastic.

The various construction types and dimensions of wire rope sockets are mainly according to the tension force to be transferred. Further criteria are: assembly conditions, requirement for retensioning, corrosion protection and maintenance.

High wire rope breaking loads can be achieved by end connections with sockets filled with metal. When destructive tests are carried out on ropes with metal-filled sockets, the rope usually fails in the free length.[1] The rope is opened out at the end to form a brush shape and cast into a conical socket. This is called speltering. With increasing load, the metal cone pulls deeper into the socket with constantly increasing clamping forces. The filling can use various materials, but today is almost always cast with zinc (Zn 99,9) or Zamak Z 610 (Zn Al6 Cu1).[2]

To perform the casting operation, the socket is checked for surface damage and alterations to the internal structure through temperature. The rope to be anchored is unwound from the cutting point to the end of the future cast cone and cut off, not with a flame cutter. Now the end of the rope and then each strand are fanned out using pieces of pipe to form a "brush". Steel inserts of ropes are also unwound and fibre inserts are cut out. The resulting rope brush is carefully cleaned and degreased with a cold degreasing compound. Bare, ungalvanised wires have to be roughened with a corrosive compound and tinned.

The rope brush is now pulled into the socket and mechanically fixed at the socket opening. The socket is hung up vertically, so that the rope enters it vertically and can be cast in stretched condition. Before filling, the socket is warmed with controllable burners in order to maintain the melting temperature of the filling material for a sufficiently long time so that the material can flow into all parts and the formation of voids is prevented. When pouring the filling, it is important that the material pours slowly and evenly into the socket, so as to avoid air bubbles.[3]

In order to ensure the force transfer from every wire into the filling, the filling operation has to be carried out very carefully.

An alternative to metal filling of the rope sockets is plastic filling. This is a very reliable rope end connection which, if made correctly, can achieve higher loads in destructive tests

Rope brush

Prewarming the socket   Filling the socket

Fig. 25: Production of a cylindrical cast socket

1 Mogk, R. (2000)
2 Gabriel, K.; Wagner R. (1992)
3 Verreet, R. (1996-1)

Fig. 26:
Forked rope socket with Zamak fillings

than can be achieved with metal fillings. The more elastic filling permits a relaxation of the wires near the breaking point and thus ensures that all wires bear the load evenly.[1]

The plastic filling material usually consists of polyester or epoxy resin, a hardener and a filler, which reduces the overheating of the cone during the curing reaction and reduces the shrinkage on cooling. To improve the compressive strength, a filler (quartz powder, steel balls) is used.

The major advantage of the filling process with artificial resins is that, in comparison to metallic filler, there is only a minor effect of heating the steel wires. Further advantages of the plastic filler are the comparatively lower weight and that it can be implemented on the construction site without special equipment. The long-term behaviour of plastic filler has not yet been sufficiently investigated for lifetimes over 10 years, which explains why it is not very commonly used as an end connection for stay wires.[2]

Swaged fittings

Open spiral ropes and round strand ropes can transfer forces to aluminium or steel sleeves pressed onto the rope permanently. The transfer of force is by friction and form-fit. Swaged sockets are produced as eccentric connections, where the rope eye (loop) lies in a thimble. Axial connections with centrally acting loading are made as Flemish eye or as swaged socket fittings, where the straight rope end is fitted with a fork or eye fitting.

The most commonly used permanent end fitting in Europe is the swaged aluminium sleeve. It can be produced easily and cheaply. Through cold forming, the aluminium is compressed so that a friction bond is created between the rope strand and the wall of the sleeve. In tension fatigue tests, aluminium swaged fittings achieve the required breaking load and a good lifetime.[3]

To protect the rope against excessive transverse compression, the rope eyes can be laid in pressed steel thimbles or solid thimbles. A relatively wide spread angle of the rope eye is required for use with a thimble. In order to avoid tearing of the swaged fitting, eye diameter, rope diameter and the spacing of swaged sleeves to the thimble need to be correctly coordinated.

To manufacture an aluminium swaged connection, the first step is to select the required form of compression sleeve.

The types available are cylindrical, cylindrical-rounded and cylindrical-tapered with a window to see the position of

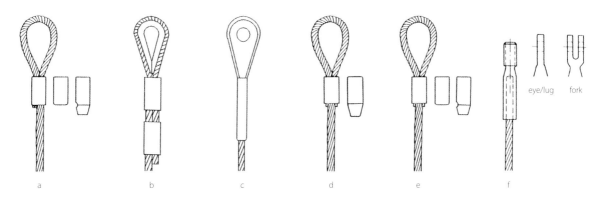

a Aluminium swaged sleeve connection
b Aluminium swaged socket with pressed steel thimble
c Aluminium swaged socket with solid thimble
d Flemish eye with steel sleeve
e Flemish eye with aluminium clamp
f Swaged socket fitting

Fig. 27: Swaged fittings

1 Beck, W. (1990)
2 Verreet, R. (1996-2)
3 Hemminger, R. (1990)

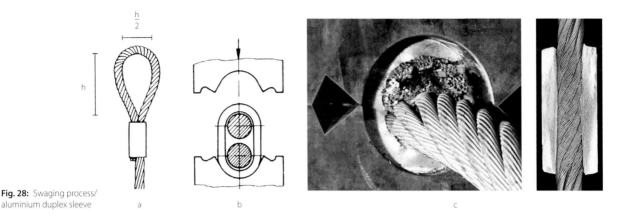

Fig. 28: Swaging process/aluminium duplex sleeve

a   b   c

the end of the rope (a, b, c in Fig. 27). The size of the swaging sleeve is chosen according to the rope diameter, the rope construction and the filling factor. After the rope has been cut to length, the end is inserted through the swaging sleeve, bent to form a loop, fitted into a thimble and back through the swaging sleeve. Then the sleeve containing the "live" (load-bearing) end and the "dead" rope strands is laid in a hydraulic or pneumatic press, aligned in the pressing direction and pressed in one pass until the sides of the die close together (b in Fig. 28). This process is called **swaging**. Projecting wire ends are filed off.[1]

After the pressing is completed, the dead rope strand must be flush with or stick out of the end of the sleeve. With tapered sleeves, the end of the rope must be flush with the cylindrical part.

After ordinary lay ropes have been swaged together, the outer wires lie alongside each other (c in Fig. 28). After Lang's lay ropes have been swaged, the outer wires are crossing and can notch each other; this normally leads to no great reduction of the breaking load.[2]

Where a slimmer style is required for smaller diameter ropes, socket fittings can be swaged onto open spiral and round strand ropes by pressing, drawing, rolling or hammering. Versions are available with fork, eye, threaded stud and ball. The advantages of this connection are slender form, the multitude of connection methods, central force transfer and ease of manufacture. The force transfer from the wire rope to the socket (terminal) is by form-fit (interlocking). Swaged socket fittings can reach rope breaking loads of 90–100 %.

For the production of swaged socket end connections in the construction industry, the most commonly used methods are pressing and rolling of fittings. In addition to pressed and rolled fittings, versions with adjustable turnbuckle are also available for rope diameters up to 20 mm, which can be used

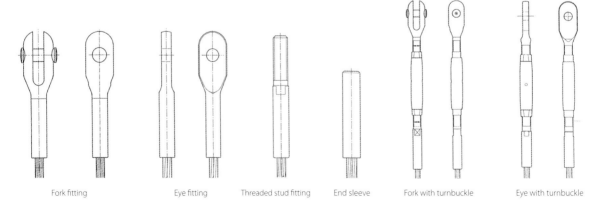

Fork fitting   Eye fitting   Threaded stud fitting   End sleeve   Fork with turnbuckle   Eye with turnbuckle

Fig. 29: Various forms of swaged socket fitting

---

1 Verreet, R. (1996-2)   2 Verreet, R. (1996-2)

Fig. 30: Swaging a socket

a Terminal rolling machine

to adjust tolerances (right in Fig. 29). For galvanised or Galfan galvanised open spiral ropes, galvanised fittings are used. Spiral ropes of high alloy steel are correspondingly anchored with high alloy fittings.[1]

To make a swaged socket end fitting, the marked end of the rope is cut off flat at a right angle and inserted into the drilled out and deburred shank of the socket. The insertion length is as a rule 4 to 6 times the rope diameter. The pressing operation is mostly done with a stationary toggle hand press or a hydraulic press (Fig. 30).

An increasingly used process is the milling or rolling of socket fittings. The socket is hydraulically drawn through a terminal rolling machine synchronised with gear wheels and pressed onto the rope bit by bit. (a, b in Fig. 31). The advantage of this process is the relatively low power requirement of the machine, with the result that rolled fittings can be made on transportable machines on the construction site. A further advantage in comparison to the pressing of sockets is that the socket is elongated during the rolling process without elongating the rope, which has a positive effect on the fatigue tension test results.[2]

If the swaging sleeve has a tapered shape at the end, this largely avoids sharp transitions and basket-like bulging of the strands (c in Fig. 31).[3]

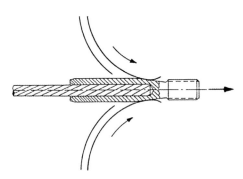

b Diagram of the rolling process

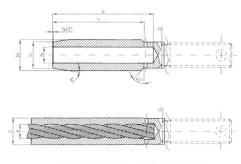

c Socket geometry before and after the shaping

Fig. 31: Rolling of sockets

1 Mogk, R. (2000)

2 Verreet, R. (1996-2)

3 Vogel, W. (2002)

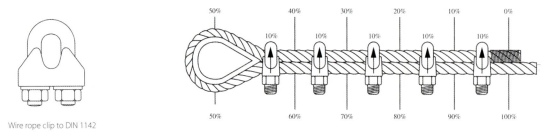

Fig. 32: Force distribution in an end connection with wire rope clips

Wire rope clips

For wire rope diameters of up to 40 mm, bolted wire rope clips can be used as an end connection. Wire rope clip end connections can be made with or without thimble. Because they can be undone easily, they are best suited for temporary construction. Compressing the "live" rope end onto the "dead" end enables force transfer through friction and form-fit. Each clip transfers about the same load, until the dead rope end is unloaded.

When no thimble is used, the loop should be made with a length of at least 15 times rope diameter, and the free spacing between the clips should be 1 – 3 times clip width.

For the assembly, it is important that the saddle is next to the loaded rope and the U-bolt is next to the dead end. The threads of the U-bolt and the contact surfaces are to be greased and the first clip should be fixed tightly next to the thimble. The nuts are to be tightened again with a torque wrench a reasonable time after the first loading.

Foundation connections, connections to other building elements

The forces, which act on tensioned wire ropes, must be transferred into an anchorage at the end of the rope, where they are transferred into other elements, mostly masts, columns or edge beams; or can be connected axially. Where structures are stayed directly to the ground, these forces must be passed into the subsoil through tension anchorages, which in most cases also function as compression member. It is normally possible to adjust tolerances.

The connection can be directly to the foundation or be made with fixing brackets. Fully locked wire ropes are normally bolted with forked sockets to a fixing bracket. The tolerance can be compensated using a mortar layer under the fixing bracket.[1]

Threaded rods are normally concreted in to hold down the fixing bracket. The fixing brackets with the bolted wire ropes can then be used for hydraulic pre-tensioning (right in Fig. 34).[2] With smaller diameters of rope, the forces can be applied through swaged fittings with adjusting threads, or turnbuckles. For large tension cables, a crosshead with threaded bars can be used for adjustment (left, middle in Fig. 34).[3]

For the tension anchorage into the subsoil, a combination of reinforced anchor block and piles can be used. The piles need to have a stiff connection at their heads. The ground

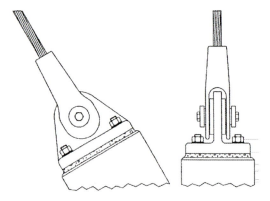

Fig. 33: End connection to foundation

[1] *Mogk, R. (2000)*

[2] *Stauske, D. (2002)*

[3] *Stauske, D. (1990)*

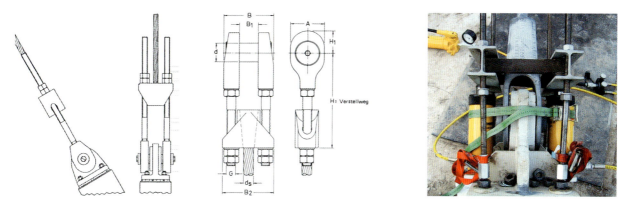

Fig. 34: *left:* Crosshead with threaded tie bars; *middle:* Tolerance adjustment for highly loaded cables; *right:* Tensioning equipment

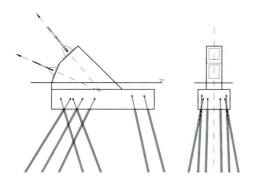

Fig. 35: Construction of elevated anchors for stay ropes

Rope anchorage to edge beam

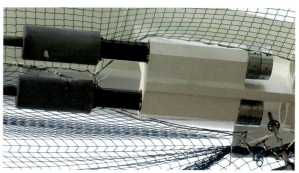

Cylindrical speltered socket internally threaded with screwed-in tie bar

Fig. 36: Connection to other building elements

anchor can be constructed below ground or elevated as required. The bearing capacity of the anchorage is essentially determined by the angle of the tension force and the location of the point of application of the load.

If the forces in wire ropes have to be transferred to other building elements, the compensation of axial tolerances can be provided by adjustable cylindrical rope sockets with packing shims, or threaded tie bars with external nuts.[1]

---

1 *Detailed descriptions of tensioning equipment and devices for ropes can be found in section 3.3.2*

2.2.1.2 Webbing

Flexible edges of textile membrane surfaces can be reinforced parallel to the edge by attaching webbing.

If the edge detail is flexible with ropes running in cable sleeves, then tangential stresses arising along the edge of the membrane could cause displacements between the rope and the membrane. For edge details where the friction forces between the rope and the cable sleeve are insufficient to handle these displacements, additional webbing can be attached. Depending on the extent of the tangential forces to be expected, these can be fully or additionally resisted by the edge webbing through the fixed connection with the textile membrane. The bonding strength between webbing and membrane is thus essential for the transfer of these forces.

The relative stiffness of the membrane and the edge webbing have to be taken into account, and the elastic properties of the webbing are mostly different from those of the membrane. Webbing normally stretches more than fabric membranes and is therefore pre-stretched before attaching to the membrane. Especially with flexible edge details, where the webbing alone resists the tangential forces under load, the edge webbing cannot be allowed to hinder the deformation of the edge of the membrane.

In addition to the reinforcement of edges, webbing slings are used as temporary equipment in the erection of wide-span lightweight structures. They are primarily used as staying ties for the temporary stabilisation of primary or substructure elements (see section 3.3.4). Webbing tie-downs are also used in combination with ratchet systems for tensioning stiff membrane edges (see section 3.3.2).

The webbing used consists of polyester or polyamide. The webbing is sewn or glued to the edge of the membrane in one or two layers. Polyester fibre webbing has a better tearing strength and clearly better shrinkage behaviour than polyamide webbing, and so is mostly used. Webbing exposed on the upper side should be protected from UV rays and dampness.

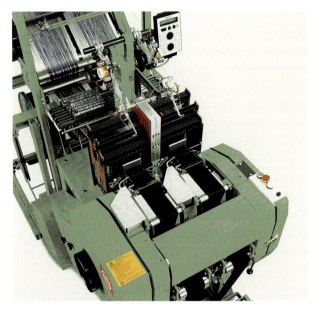

Fig. 38: *top:* Weaving system with simultaneous weft insertion
bottom: Needle weaving machine

Fig. 37: Polyamide and polyester webbing

Fig. 39: Webbing fixing to corner fitting

Both types of webbing are manufactured in weaving works. The normally Z-shaped yarn is woven on needle looms. This type of automatic weaving machine runs at high speed and the weft is inserted from both sides simultaneously by needle holders. The webbing is woven in plain or twill weave.

If required, the yarn can be impregnated to make it water-resistant or flame retarding. Mould can be largely prevented with a fungicide treatment. To increase the pliability, lubricant can be added. For applications requiring high tensile strength, aramide webbing can also be used.

After weaving, the webbing is delivered for fabrication on reels, where it is pre-stretched according to the specified compensation, cut to length and sewed under tension to the membrane edge, ensuring shear-resistance.

Depending on the demands of assembly and the application of pre-tensioning, webbing can either be anchored to the end mountings with fitted keders over brackets or pulled through openings in the mountings (Fig. 39).

At corner details where the membrane is pulled toward a mounting, which is held in position, the webbing belts can be tensioned with belt tensioners through intermediate links. Holed and formed metal plates are used as intermediate links.

The arrangement of belt tensioners (Fig. 41) is useful during erection, particularly for precise tensioning or detensioning of heavy fabrics.

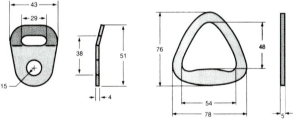

Fig. 40: Hole and flat triangle as intermediate link

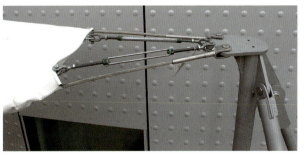

Fig. 41: *left:* Corner detail with belt tensioners; *right:* tensioning the edge webbing

2.2.1.3 Keders

To form a stiff edge detail for membranes, the pieces are held at the edges by metal profiles (clamping plates, keder rails). The force transfer from the membrane to the metal profile is ensured by a cord-shaped keder trapped in the metal plate.

The purpose of the keder is to prevent the membrane from sliding out of the profile (bottom in Fig. 42). The keder acts as a linear bearing element in a connection, resisting forces through form-fit. It must therefore be designed for the relevant tension. The diameter of keder cords is 5–12 mm.

Keders made of PVC monofilaments, polypropylene or polyurethane are mostly used in textile construction. The keder is manufactured with a round section in an extrusion process, calendered with special orifices and then wound on reels.

The materiel used for the keders depends on the material and weave structure of the fabric to be used. When selecting the material, the differing stretch properties of the membrane and the keder have to be taken into account. Keders with varying Shore hardness are used depending on the required stretch properties, with the usual value being SH 50–SH 90. PU keders have a higher wear resistance than PVC keders and are coated to be UV resistant.

When using lighter types of membrane (type 1– type 2), the fringed keder can be delivered prefabricated to the fabrication as keder with flap and welded to the membrane. When using heavier membranes, the membrane is folded over the keder during fabrication and welded.

Fig. 43: Welded-in keder in a PES/PVC fabric type 5

When assembling edge details with one-part keder rails, the keder cord has to be pulled through the rail during erection. In order to ensure that it slides properly, the keder seam should be tidily welded. Keder cords with a low Shore hardness warm up with friction more than harder cords. In order to make sure that the membrane slides easily into the keder profile, a lubricant is often used.

Depending on the edge geometry, erection method and extent of the forces to be transferred, steel rope or aluminium keders can be detailed as an alternative to plastic keders.

2.2.2 Surface load-bearing elements

Wide-span, light surface structures made of materials, which serve both structural purposes and provide protection against environmental influences, are normally called *membrane construction*. This description (lat. membrana = skin) implies on the one hand a function and on the other a dimension, and thus signifies a skin-like thin material. Such materials in construction are commonly called technical membranes. This group includes technical textiles (coated and uncoated fabric) and also technical plastics (extruded films).

Technical membranes, as used in construction for wide-span light surface structures, are materials consisting of elements of large area joined together, enabling load transfer exclusively in tension. To do this, they have to be curved, edged and anchored in a suitable manner.

Large spans and the resistance against the effects of load, time and temperature require technical membranes with the appropriate composition for the purpose. If this composition is to be achieved through the arrangement of single elements or layers, then this is called *bonded construction* or *composite materials*. Coated textile fabrics are such a composite material. They consist essentially of artificially manufactured materials (plastics) formed by the linking or modification of molecules on an atomic scale.

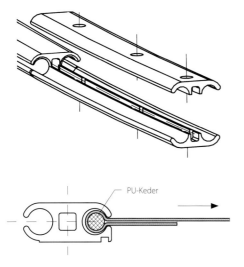

Fig. 42: Butt joint with one and two-part keder rails

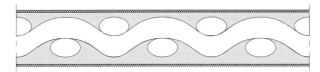

Textile fabric composition

Thermoplastic film

Fig. 44: Diagrammatical section through a planar element

Coated textile fabrics consist of three layers of synthetic polymers. They fulfil the requirements stated above for wide-span surface structures and are the most commonly used group of materials in textile construction (left in Fig. 44).

A second group of materials are technical plastics made of fluorothermoplastics in the form of film. Because they are nearly transparent, these single-layer fluorothermoplastic films have gained increasing popularity among architects. They do, however, have the disadvantage compared to coated fabrics that their relatively lower strength can only be used for smaller spans. As a result, they have to be supported along their edges by stiff or flexible primary or secondary structural elements. The surface is as a rule detailed with more than one layer supported by air, which results in considerably more advantageous thermal insulation properties than with textile fabrics.

Experience gained with conventional materials and methods of construction is insufficient to describe the material behaviour of large-area flexible elements or the practical implementation of membrane construction, whose mechanical properties show essential differences from those of traditional materials. The following section therefore attempts to outline the composition, the types and the manufacture of the sheet materials used in membrane construction.

One important aspect of building with form-active structural systems is the deformation behaviour of the materials used, so the most important mechanical properties of the flexible sheeting will be summarised here. Patterning, the determination of the cutting patterns for the arrangement and jointing of the flexible pieces of fabric to form the structural elements, has a major importance in the process of practical implementation. The factors influencing this process will therefore be described here in more detail. The different details to overcome the problem of force transfer through the edging are also explained.

Coated fabrics are the most commonly used materials for wide-span lightweight structures, and so are treated here in the most detail.

2.2.2.1 Coated fabrics

Textile fabrics form a system from woven yarns, which are arranged orthogonally when unstressed and consist of single threads, parallel or twisted together. They have been used in many fields for hundreds of years and enjoyed a renaissance in the construction industry during the last century among committed architects and engineers, particularly the architect Frei Otto.

The symmetrically structured raw fabric is treated with additional compounds, coated with special pastes and the surface is treated to protect it from external influences. Two types of material are mostly used for coated fabrics in construction: polyester fabrics with a PVC coating and glass fibre high-strength fabric with PTFE coating. Silicone-coated glass fibre fabrics are also in use. Fluoropolymers and polyolefin-coated polyester fabrics, coated fluoropolymer fabrics and PVC-coated aramide fabrics are less commonly used. Depending on the requirements resulting from the forces to be transferred, a material group is specified according to the type of fibre, weaving and coating.

For PVC-coated polyester fabrics, synthetic polymer fibres (Polyethylene terephthalate = PET)[1] are the most commonly used, the fineness (yarn count) of which is expressed in dtex.[2] The fibre has a considerable influence on the material properties of a fabric. The major advantage of PET fibres is their dimensional stability against the effect of chemical and physical influences and that they don't rot. PET belongs to the group of partially crystalline thermoplastics and the melting point is 265 °C.[3] Above this temperature, the material becomes soft and has viscous properties.

---

[1] The description PES fabric used by fabricators when used for wide-span lightweight structures derives from the internationally agreed short form PES for polyester, and should not be confused with the abbreviation PES for polyether sulfone resin, which belongs to the group of polysulfones. Chemical nomenclature uses the abbreviation PET for polyester.

[2] 1 decitex = weight in grammes per 10,000 m of yarn, 10 dtex = 1 tex

[3] Fritz, C. P. (1999)

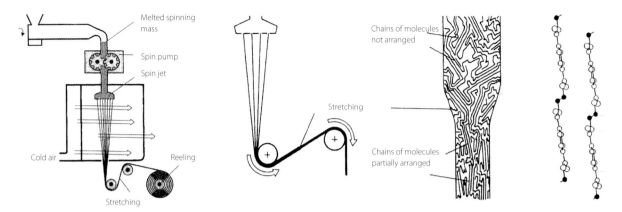

Fig. 45: Diagram – *left:* Spinning process; *middle:* Stretching; *right:* Crystalline structure of PET polymer chains

The decisive factors determining the strength of fibres are the orientation and bonding of the chains of molecules with each other and their degree of polymerisation (chain length). PET fibre polymer chains are arranged in triclinic symmetry (right in Fig. 45). Triclinic crystal systems have the least symmetry, no equal angles and no axes of the same length. In order to improve the mechanical properties of the fibres, further processing is necessary to produce an elongated, nearly parallel arrangement of the molecules. To achieve this, the monomers are polymerised, the resulting granulate is spun to fibres and then evenly stretched in a stretching process.

After the production of the starting material, the granulate is melted in a melting and spinning process, filtered and led to a heated spinning manifold.

Then the spinning mass at a processing temperature of 280–300° is pressed under pressure through jets into a spinning shaft and formed into filaments. After leaving the jet, the liquid filament is fixed by uniform blowing of air. In this condition, the macromolecules of the fibre are pre-oriented. The orientation parallel to the fibre axis is produced in the stretching process, where the raw, already solid fibre is fed through a variable-speed roller system and elongated to many times the original length (left in Fig. 45).

Finally, the fibres are twisted together. This can be either Z-twist (right-hand) or S-twist (left-hand). The sum of the fibres spun out of the jets and twisted is called yarn.

Types of fabric making

A raw fabric is made by the interlacing of the warp and the weft threads, and the thread density gives the number of threads in warp and weft. Various methods of crossing the warp and weft threads lead to the characteristic waviness (crimp) of the three-dimensional fabric. Two types of weaving are used to make fabrics for membrane construction; plain weave, the simplest and closest crossing of warp and weft, and basket weave, a modified plain weave.

With both types of weave, the warp runs in the long direction of the fabric. Because of the weaving process and the coating operation, the warp and weft threads have different geometries. While the warp thread, which is kept under tension, has a more

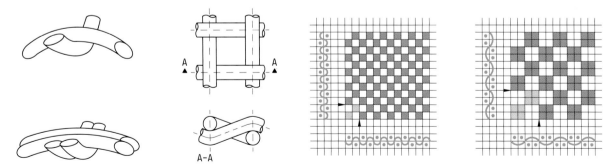

Fig. 46: Diagram of the weave of the main weave types with warp and weft section

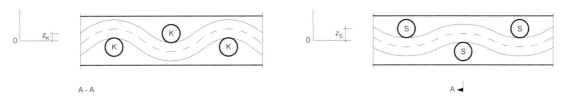

Fig. 47: Diagram showing section of weave with differing amplitude displacements in warp (K) and weft (S) sections

stretched curve, the weft shows stronger crimping in the bedded coating (left in Fig. 47). As a result of this spring in the weft, the weft thread is softer than the warp and has lower moduli.[1]

Weaving

Most types of polyester high-strength fabric are woven on a rapier loom. The parallel or twisted bundle of warp threads is unrolled from the warp beam and fed under tension into the automatic loom, where the weft thread is between the opened layers of warp (shed) by a rigid rapier. The weft thread is inserted into the open shed by the left rapier and passed by the rapier clamp in the middle of the fabric to the right rapier, which takes the thread to the right-hand edge. The tension in the weft thread is controlled by the built-in thread brake. At the change of location of the warp thread, the weft thread receives the typical wave shape.

Fabrics are woven in widths of up to approx. 5 m. In Austria, fabrics can be made at a maximum 3.2 m wide. The normal roll widths are 2.05 and 2.50 m. Depending on the fabric width and the area weight of the raw fabric, roll lengths of a maximum of 2,200 running m are usual, with increasing surface weight underneath.

According to the weave type, thread thickness and roll width, weaving by the rapier weaving process lasts 1–3 days per 1,000 running m and is the most time-intensive process in the manufacture of fabrics.

Measures to accelerate the production process are still in development. The application of air weaving, where the raw thread is stretched by supporting air, is being investigated. The advantage of weaving with air-jet looms is the increase of production speed; it is possible to insert more wefts per minute.

Coating, treating, rolling

Textile fabrics are subject to numerous factors influencing the ageing process and thus the quality requirements. In addition to load-dependant influences like alternating loads, extent of loading and creep, other factors and above all load-independent effects like ageing, natural climatic and atmospheric effects caused by pollutants have to be taken into account with textile fabrics. Textile fabrics, in addition to undertaking structural functions, have to be resistant against chemical and biological influences and also be almost non-flammable.

Fig. 48: *left:* Automatic rapier loom; *right:* Rapier clamp with weft thread

1 Blum, R. (2000)

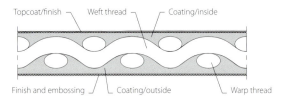

Fig. 49: Internal construction of a polyester fabric coated on both sides and with topcoating

The resistance of materials against external effects is achieved through a wide range of measures during production. Textile fabrics used in construction for wide-span lightweight structures are therefore designed for specific purposes as composite materials. With various combinations and thickness of materials for base fabric, coating and finish, various deleterious effects can be resisted. With coated membranes, the fabric undertakes the load bearing function. The coating protects the fabric from damage. Together with the surface treatment (topcoating, finishing), it ensures the sealing of water and air.[1]

The coating material for polyester fabrics is PVC-P (Polyvinylchloride-Plastisol with plasticizer and additives), a thermoplastic made of polymers with a linear, unbranched molecule chain arrangement. With the application of heat, it can be formed and becomes flowing. Coatings are applied to PVC-coated polyester fabrics in order to delay the escape of plasticizer from the coating and thus protect the fabric coating from becoming brittle. If the protection against UV rays failed, this would lead to a rapid decrease of strength in the short term.

This surface treatment also hinders the growth of sedimentation and the formation of microbes with the resulting impairment of the optical properties.

PVC-coated polyester fabrics are mostly coated using a knife coating system in Europe. An exception is the process for coating pretensioned polyester fabrics in the Precontraint process (Ferrari). In a process used in the USA (Seaman process), the weft thread is laid on the warp thread and then the coating compound is laid on top.

In the painting process, the first stage is to apply the adhesion coat on the raw fabric. The adhesion coat is applied in the form of a paste with an air knife. An adhesion compound (containing isocyanate) with the addition of plastisol (PVC-P, amorphous thermoplastic) guarantees the necessary bond between the raw fabric and the coating (adhesion strength) for subsequent processing. The adhesion coat is heated to approx 180 °C in a fusion tunnel. This makes the plasticizer penetrate into the PVC granules very quickly, swell them up and fuse them together. In the jelly-like soft PVC layer, the PVC and plasticizer molecules combine to a homogeneous mass. In order to be able to unroll the coated fabric leaving the fusion tunnel, it is cooled by passing through a number of water-cooled rollers.

After cooling, the adhesion layer lies like a film on the raw fabric. In a second process, the top coat is applied with an air knife. This contains the same components as the adhesion layer, but without adhesion compound. This is followed by another partial gelling and cooling. Because textile fabrics are normally coated on both sides, the entire process has to be repeated for the reverse side.

One criterion for the long-term preservation of the mechanical properties of polyester/PVC fabrics is the coating thickness on the thread ridge. The fabric protection is directly proportional to the coating thickness: "The higher the thickness of the covering, the better the protection of the thread."[2] The usual coating thickness is from 0.08 mm to 0.25 mm. For fabric type 1 (plain weave) with a total coating weight of 500 g/m², 200 g/m² are applied to the fabric underside

Fig. 50: Polyester fabric exposed through breakdown of the coating

1 *Fluoroplastics films, which are primarily used in pillow construction, must in contrast fulfil all functions.*

2 *Blum, R. (2002-1)*

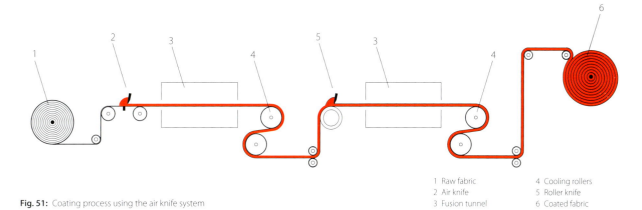

Fig. 51: Coating process using the air knife system

1 Raw fabric
2 Air knife
3 Fusion tunnel
4 Cooling rollers
5 Roller knife
6 Coated fabric

and 300 g/m² to the upper side. The base coat on each side of the fabric is approx. 100 g/m². The remaining weight is represented by the top coat. For raw fabrics with higher area weights (basket weave), the base coat layer is about 200 g/m². The number of layers of coating depends on the fabric type. Depending to the number of layers, the coating speed is approx. 10–15 m/min.[1]

Some manufactures possess a "4-coat" coating plant, where the PVC coating is applied to both sides of the fabric in one production pass. The first three layers are applied by vertical coating equipment and the last coat, normally the top coat of the upper side, is applied with a horizontal roller knife.

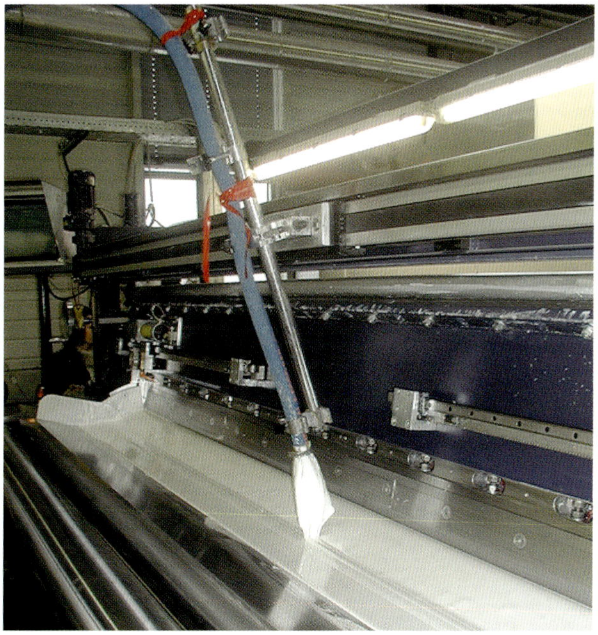

During the production steps of weaving and coating, the warp is held under tension. The coating plants are controlled so that a uniform tension can be applied to the warp. Along the edge, the fabric acting as substrate for the coating is held in the tension frame with needles or clips during the gelling process. Because of the shrinking forces in weft direction, there is the danger of the needles not holding, which is why pre-shrunk thread is used, with a shrinkage in hot air of max. 2.5 %.[2]

The surface treatment of the coating, called topcoating or finishing, gives an additional protection against the fabric becoming dirty too rapidly or moisture penetrating. The treatment also delays the loss of plasticiser from the coating.

Two processes are available for the application of the top coating, which affect the properties of the finished membrane in different ways: the lamination of films and the application of paints. Lamination achieves protection against external influences through the application of a 0.03–0.08 mm thick, solid and transparent film to the PVC coating. An absorber contained in the primer makes the film resistant against UV effects and a white pigment additionally shuts out and reflects the UV light. This technique does, however, have disadvantages for the manageability of the fabric. The lamination with film additionally hinders the flexibility of the warp and weft threads at an angle (see section 2.4.1).

When applying paint based on PVDF (fluoroplastic) and acrylates, the fabric is only negligibly stiffened. The paints, which are applied from solvent systems, are applied to the inner and outer sides in a layer thickness of approx. 5–10 μm. PVDF paints can be non-weldable paints, which have to be scraped off before welding, or else weldable derivatives made of PVDF paints and acrylates. PVDF paint is applied to the outer side with at least 2 coats. Top coat coatings consist-

Fig. 52: Top coat with roller knife

1 Graf, W. (2004)

2 Graf, W. (2004)

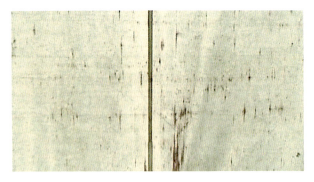 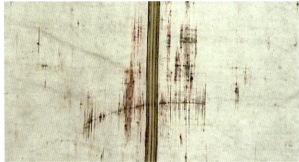

Fig. 53: Wicking effect on polyester/PVC fabrics

ing of PVDF paints ensure a surface of unaltered appearance in the long term, but acrylic paints cannot prevent discolouration in the long term.

The importance of treating coated fabrics can be seen from the various effects resulting from moisture, which can affect fabrics with damaged coating.

One example of this is the wicking effect, a by-product of osmotic pressure. At damaged locations in the coating, water can be transported along the thread and damage the fabric permanently (Fig. 53).

In addition to the alterations caused by the effects of temperature, light and natural media, material weakening through attack by microorganisms and chemicals also has to be expected. Fungi, lichens and the tips of roots exude metabolic products in the form of acids (oxalic acid, sulphuric acid), enzymes and dyes, and these can can alter the properties of the material permanently. The biogenic alteration caused in this way can lead to the coating coming loose (Fig. 54). With

Glass/PTFE fabrics, the good anti-adhesive capability of the PTFE coating on both sides has the effect of preventing deposits on the coating.

In the last step of the process, the coated and treated fabric is rolled on the outer side. This process produces a flatter surface with the result that the fabric stays cleaner. The rolling is done with steel rollers, which press the fabric against rubber-faced opposing rollers.

If there are special requirements for durability or fire behaviour, then one of the group of PTFE-coated glass fabrics will be used.

The stiffness and strength of these fabrics comes mostly from the glass fabric. The protective PTFE coating consists of fluorinated plastic, one of the strongest bonds in organic chemistry. PTFE is a less stiff, but very strong plastic. It is to a high degree resistant to chemicals, temperature, light and weather-resistant and has very good anti-adhesive properties. PTFE is also, without the addition of stabilisers and plasticisers,

Fig. 54: Separation of coating from a polyester/PVC fabric caused by fungal attack

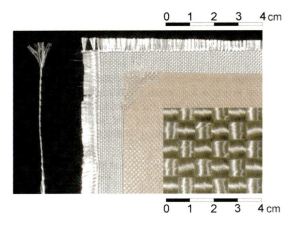

Section in warp direction

Section in weft direction

Fig. 55: *top:* Glass fibre, glass fabric and coated fabric; *bottom:* Internal construction of fabric

non-flammable and can be used over a temperature range of –270 °C to +300 °C.[1] The glass fibres forming the base fabric are UV-resistant. If the coating is damaged, the penetration of water is mainly a problem on account of the loss of strength.

Glass fibres, because they are brittle, are spun in single filaments, which are then twisted together to form single threads. During the spinning and twisting together to form yarn, the threads are dipped in a size. This reduces the mechanical loading and the resistance of the filaments and acts as lubrication.[2]

The glass fabric is coated in 5 m wide rolls in 6–10 passes through a watery PTFE emulsion and then subjected to infrared rays, which fully evaporate the water. After the first stage of coating, the PTFE has to be sintered at 370–380 °C on account of its low flowing ability after fusing, which causes a part of the applied material to caramelise. This process is repeated until the coating has the desired thickness. The sintering on of PTFE particles in layers results in a fissured surface, which can be levelled with a finish of FEP. The warp direction is tensioned with a defined force during the entire coating process. In the weft direction, it is not possible to hold the fabric fast on account of the strong heating during sintering. The crimping of the weft thread resulting from the weaving remains and is made still stronger by the tensioning in the warp direction.

After completion of the coating process, the fabric is cooled. The different coefficients of thermal expansion of glass fabric and PTFE and the more rapid cooling of the outer layers lead to the PTFE expanding before the glass fabric. The crystal transition at 19 °C is reinforced by this effect at low processing temperatures.

Because of the caramelisation during coating, the coating initially discolours to light brown. This colouration disappears through bleaching after a few months of UV exposure. Apart from this colour change from brown to white, there are no appreciable effects on Glass/PTFE membranes from long-term radiation intensity.[3]

One disadvantage of PTFE-coated fabrics is the susceptibility to kinking, which requires a special edge detail. The preparation work for erection (packing and transport) and the erection have to include suitable measures to prevent the occurrence of breakages in the coating. Such breakages could be the starting point for further tearing. These fabrics are therefore not suitable as structural elements in temporary buildings and convertible construction. In addition, they are nearly rigid in shear and hard, are difficult to handle and can only be erected without damage at temperatures over 5 °C. On account of the strong creep, considerably more time is required for the erection.

A group of materials, which are seldom used, are silicone-coated glass fibre plastics. The carrier material for this fabric is interwoven filaments of silicate glass. The coating material is a clear to opaque silicone, treated with additives.

The advantage of these fabrics is being especially flexible and not prone to kinking. They fulfil all strength requirements and have favourable tearing behaviour. They are largely resistant against chemical effects and can be used at temperatures from -60 ° to +180 °C. They are also UV resistant, do not become brittle and can be coloured as required. One disadvantage that should be mentioned is the behaviour of the coating in getting dirty. Silicones are resistant against chemical attack, but their surface charges up statically and attracts dirt.[4]

To prepare it, the silicone rubber has to be kneaded, dissolved in solvent and mixed. The silicone is applied to the

---

1 *Oberbach, K. (2001)*

2 *Baumann, Th. (2002)*

3 *Baumann, Th. (2002)*

4 *Blum, R. (2002-2)*

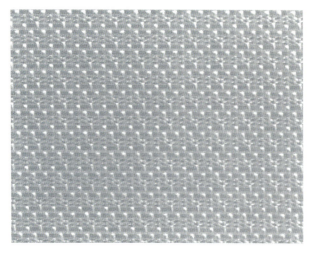  

glass fabric with a knife coating system. The silicone mixture in paste form is applied to both sides of the glass fabric with many passes of an air knife and then a top coat of silicone mixture is applied. Weathering tests are constantly carried out by the manufacturers to investigate and evaluate the behaviour with regard to attracting dirt.

Silicone cannot be welded thermally. It has no pronounced melting point and hardens to an elastomer with spatially networked molecules. The roll material up to 2 m wide is, according to information from the manufacturers, sewed, stapled or vulcanised.

The vulcanisation of the strips is carried out with beam presses. A silicone adhesive tape is laid between the parts to be welded and heated to 150–170 °C. The resulting vulcanisation reaction between the adhesive made of synthetic polymers with organic silicon compounds and the rubber molecules of the silicone coating leads to a high-strength joint. The welding duration with this process is between approx. 30 sec and 2 min.[1] With this jointing method, care has to be taken with the temperature resistance of the glued seam. In order to achieve form stability and freedom from shrinkage, the product has to be stored at particular temperatures for a particular time, which ensures mould striking by the end of the reaction.

Fig. 56: *top left:* Silicone-coated glass fabric
top right, below: Production

1 Funk, J. (2005)

2.2.2.2 Foils

The economic importance of plastic foils has increased enormously since the middle of the last century. They still offer today rational and increasingly ecological solutions for technical applications. For weather-resistant use in architecture, the types most commonly used are transparent fluoroplastic foils. This thin sheeting material is available after production and processing in roll form.

Fluoroplastic foils are manufactured as flat or blown films. The thickness of the flat films is between 50 and approx. 250 µm, and for blown films between 50 and 150 µm. Flat films are available in widths from 1.5 – 2.2 m, blown films up to maximum 1.7 m (cut out 3.4 m). The flexibility of the films depends on the thickness of the material from the roll and on the type of raw material used. They are UV-stable in the long term, UV-transparent and suffer no noticeable ageing.

The foils used for lightweight structures are heat-resistant, partially crystalline fluorothermoplastic foils like ETFE (ethylene tetrafluoroethylene copolymer) with a melting range from 265 – 275 °C, or THV (tetrafluoroethylene-hexafluoropropylene-vinylidefluoride-terpolymer), developed in 1983 by Hoechst, which melts in a temperature range from 160 – 185 °C (Hostafon® TFB).[1] ETFE was introduced in 1972 by DuPont (Tefzel®) in the USA and in 1974 by Hoechst (Hostafon® ET) in the European market. Later followed Asahi Glass (Afon® COP) and Daikin (Neofon® ETFE) in the Japanese market as well as Nowofol (Nowofon® ET) in Germany. The next generation of foils in the construction industry could be made of EFEP, as is offered by Daikin (Neofon® EFEP). EFEP is an ETFE strongly modified with hexafluoropropylene, with a melting range of 180 – 220 °C; the extruded EFEP foils are practically clear as glass.[2]

Chemical composition, physical structure, polymerisation
Thermoplastics are especially arranged at molecular level. The organisational arrangement (spatial layout) of their molecule chains is amorphous or partially crystalline. The molecule chains can be unbranched or strongly branched. Strong-

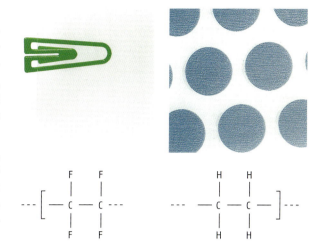

Fig. 58: *top:* Clear and printed pattern on an ETFE foil
bottom: Tetrafluoroethylene / ethylene copolymer (E/TFE)

ly branched chains increase the separation of the chains and increase the light transmission (transparency). Partially crystalline thermoplastics are opaque, and the reason for their opacity lies in the crystal size; the transparency reduces with inreasing crystallinity. Most synthetic thermoplastics are produced by polymerisation. This chemical reaction process is initiated by temperature, pressure and catalysts, and leads to a multiplication of small molecules (monomers) to larger molecules (macromolecules). The atomic bonding (primary force) of the carbon atoms forms the basis for the polymerisation of macromolecules. The coupling of similar monomers results in a polymer, the coupling of dissimilar monomers results in a copolymer. Thermoplastics are copolymers, which result from the copolymerisation of linear macromolecules.

The chain length of polymers is given in degree of polymerisation. This corresponds to the number of monomer building blocks united into a polymer chain. The degree of polymerisation of the linear and branched macromolecules is from $10^{-6} – 10^{-3}$ mm ^{3}.

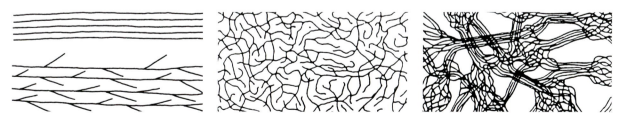

Fig. 57: *left:* Linear (a) and branched (b) chain molecules; *middle:* Amorphous (unarranged) condition; *right:* Partially crystalline

1 *Domininghaus, H. (1992)*
2 *Fitz, H. (1989)*
3 *Domininghaus, H. (1992)*

ETFE is a copolymer in thermoplastic form, whose molecular structure is reached through copolymerisation of dissimilar monomers. It consists of 50 % ethylene and 50 % tetrafluoroethylene monomer constituents. By adding a third constituent C (3–5 %) as modification, the crystallinity and the crystal size are also influenced. Without this constituent, a perfluorised vinyl compound, ETFE would be undesirably at risk of tearing under stress, especially at temperatures of 150–200 °C. ETFE is partially crystalline and the crystallinity of the raw polymer is 50–60 %. Almost complete transparency of ETFE can be achieved by a high rate of cooling.[1]

ETFE is mostly polymerised in watery liquor. During the crystallisation, contaminations are settled by stirring with the addition of a precipitant (not solvent). Finally, the dispersion is washed and dried.[2]

Preparation, pelletising

The process stages required for the manufacture of a plastic forming mass suitable for processing from the raw polymer is described as *preparation*. These processes contain the breaking up (milling, granulating) and the mixing in solid and plastic form.[3]

The plastic is prepared by addition of additives and homogenisation. To do this, it is mixed, which may be cold mixing (at room temperature) or hot mixing (at 140 °C). At this temperature, certain additives melt and diffuse into the plastic. After preparation, the plastic is in a form suitable for processing. It is first plasticised and then granulated. After the breaking up, mixing and plasticising (through heating) at 265–285 °C, the powder-form ETFE is granulated, i.e. cut into pourable pellets.

The granulation is done in a hot die-face process, in which the prepared melted mass is fed to a die plate, and exudes uniformly out of the dies. After exuding, the strings are cut off by a rotating knife roller. Water fed into the pelletiser hood is aerated to an air-water mixture and additional water is sprayed in. The cut-off granulate (pellets) is cooled in this water mist, carried out together with the granulating water and led through a cooling line to the dryer.[4]

Thermoplastic processing, extrusion and extrusion tools

The thermoplastic processing of the powder or pellet-form plastic forming mass through its viscous melted form to film in rolls is called *extrusion*. The extrusion of ETFE films is mostly done by one of two processes. In *blown film extrusion*, the melted mass coming out of the extruder is formed by a ring die into a tube, which is expanded by blowing in air. If the film in roll form coming from the extrusion is flat, then this is called *flat extrusion*.

According to the viewpoint of production quality, flat foils have advantages in transparency, shine, crystallinity, stiffness,

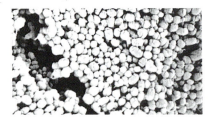

Fig. 59: *left:* ETFE granulate pellets (100–200 mm); *right:* Granulate particle

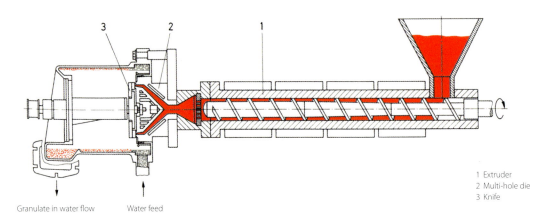

1 Extruder
2 Multi-hole die
3 Knife

Fig. 60: Hot die-face pelletising

1 *Fitz, H. (2004)*
2 *Fitz, H. (2004)*
3 *Schwarz, O.; Ebeling, F.-W.; Furth, B. (1999)*
4 *Herrmann, H. (1986)*

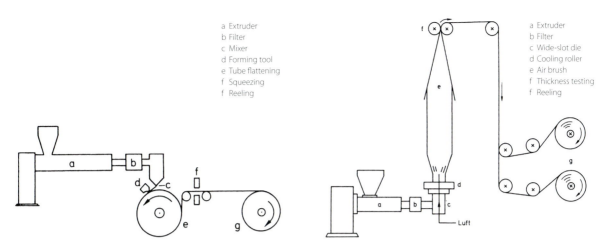

Fig. 61: *left:* Flat film extrusion; *right:* Blown film extrusion

homogeneity in the long and cross directions and thickness tolerance, while blown films have advantages regarding strength in both directions and the simple alteration of the foil width, enabled by variation of the blowing conditions.[1] A considerable disadvantage of blown films is the risk of cracking at the fold.

One of the most important devices in thermoplastic processing is the extruder. This continuously melts the pellets and conveys them to the outlets (Fig. 62). Then the melted mass is extruded from the extruder through a shaped die.[2] In flat film extrusion, only wide-slot dies are used, which form the homogenised melted mass into a flat shape.

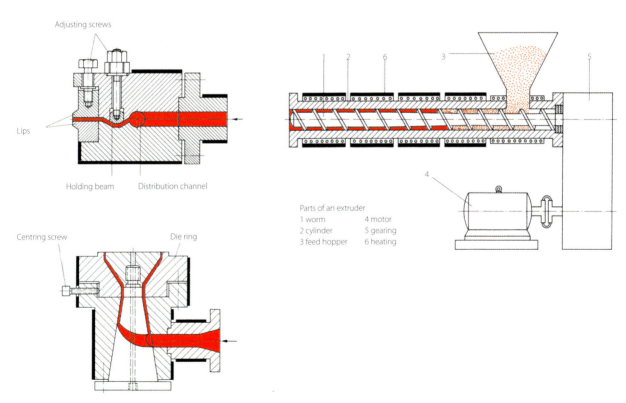

Fig. 62: *left top:* Wide slot tool; *left bottom:* Blowing head; *right:* Diagram of a worm extruder

1   Herrmann, H. (1986)   *2   Nentwing, J. (2000)*

The difficulties of designing the wide-slot tool arise from the problem of obtaining a uniform flowing front for the melted mass over the entire width.[1]

In the blown film extrusion plant, the melted mass is turned through an angle of 90° and emerges upwards or downwards as a tube. In addition to blowing head tools with radial flow, ones with central flow are also used.

The necessary process stages are shown in the diagram below and the plant components and relevant influential factors are arranged in flow direction.

The film material extruded from the plant described above is cooled by contact with rollers, stabilised, cut off and rolled up.

Fig. 64: Flat film extrusion plant

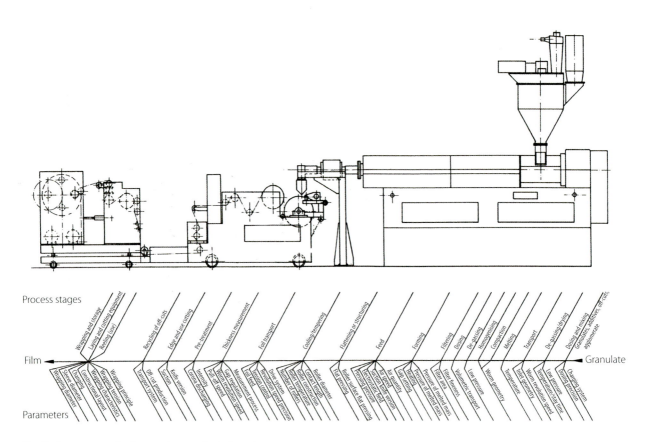

Fig. 63: Process stages – flat film extrusion according to Bogaerts

---

1 *Schwarz, O.; Ebeling, F.-W.; Furth, B. (1999)*

2.3 Material behaviour of coated fabrics

Complex demands are placed on the materials in flexible structural elements in the various fields of application. In the construction of wide-span lightweight structures, this means, in addition to the linear elements like wire ropes and webbing, above all the textile sheeting, which not only has to undertake a structural function but is also subject to load-independent effects like the ageing process and has to resist natural, climatic and atmospheric influences. They have to protect against the weather, be resistant against chemical and biological influences and be almost inflammable in case of fire. Various production measures serve to ensure the resistance of the material against load-independent effects. The most important measures for linear and surface elements were described in section 2.2.2.

In traditional construction with conventional stiff and rigid materials, the individual load-bearing elements are designed to be adequate in the most unfavourably loaded location and mostly, for practical production reasons, have the same section and stiffness over the entire length. If the bending stiffness is too low, this results in excessive deflections. The safety concepts are designed so that only negligible deflections are caused under loading.[1]

For structures loaded in tension, because the load is transferred by axial forces, it is possible to utilise the materials more efficiently and construct lightweight structures, which reflect the force transfer. For the load-bearing elements of such a structure, a certain extent of yielding under load is very advantageous for the structure. From the structural point of view, the structures are designed so that loading

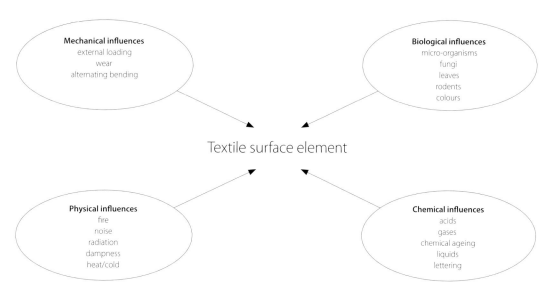

Fig. 65: Categories of long and short-term effects on PVC-coated fabric according to Dürr

2.3.1 Mechanical properties

In order to characterise the deformation of coated fabrics under load, the most important factors are the mechanical behaviour of the materials under the effects of loading, time, and temperature. Because coated fabrics are plastic composites constructed for a purpose, their mechanical behaviour cannot easily be compared with the mechanical behaviour of traditional materials. Fabric threads and coating have differing stiffness as individual elements and are subject to differing stresses causing differing deformations. Altogether, they form a statically indeterminate system.

can be evaded and deformations can be accepted.[2] The deformation of flexible elements helps to relax peaks of stress through the flexible material. A larger deformability (ductility) of the material allows the expectation of smaller deformations in the system under certain circumstances.

This contrast to traditional materials shows most clearly in the elastic behaviour of synthetic plastics. Conventional construction materials, with few exceptions, show an almost elastic relationship between stress and strain. Because of the chemical composition and physical structure of polymers, deformations arising under loading, depending on temper-

---

1 Schlaich, J.; Wagner, A. (1988)

2 Blum, R. (1990)

ature range, are partly of an elastic (reversible) and partly of a viscous-plastic (irreversible) nature. This has the result that important mechanical properties not only depend on the type of loading and the temperature, but above all on the time, the duration of the loading and the speed of loading.[1] Behaviours under mechanical loading and dependant on time, like creep and relaxation, are considerably more typical for plastic fabrics than for other materials. As a consequence, any description of the material properties and their values must also include the time scale.

The fact that the deformation depends on the structuring of the individual elements regarding their alignment in the entirety also has the greatest significance for the description of fabric behaviour. This dependence on alignment results in different deformations under load along, across and diagonally to the fibre orientation.

The interaction of all these properties, and additional effects like the temperature dependence or the behaviour of tears extending, fall outside the scope of conventional material properties. They are, however, typical for the material group of fabric membranes and demand an entire range of geometrical and structural analyses on one and two-axis fabric strips in short and long-term tests

The most important material properties for the description of the physical, geometrical and material characteristics of flexible structural elements under mechanical loading are elasticity behaviour, strength and stiffness, the tendency of tears to extend, the resistance to kinking and creep and relaxation. Questions of stability, oscillation and structural safety of lightweight structures cannot be discussed here.

PVC-coated polyester fabrics have quantitatively the largest representation in textile construction. This section is therefore primarily devoted to the load, time and temperature dependant mechanical properties of flexible materials in the group of polyester/PVC membranes, which are most relevant to practical applications in construction.

2.3.1.1 Elasticity behaviour
Non-linear and inelastic behaviour of fabrics
The deformation behaviour of a synthetic fabric can basically be described as non-linear, because it cannot be described using Hooke's law of elasticity. As a result of this, there is no linear relationship between the load applied to the fabric and the resulting strains. This can be shown in a uniaxial tension test (left in Fig. 66). The behaviour of the fabric can be described using a non-linear mathematical equation.

If a uniaxial test is performed on a fabric strip, it can be established that the resistance of the material to deformation grows with increasing action of force and that the relaxation curve differs from the loading curve. If this test is repeated, then it can be recognised that all subsequent loading and unloading curves are different from all previous loading and unloading curves (right in Fig. 66). Moreover, after loading and unloading there are permanent strains in the material, the extent of which depends on the loading history. The fabric behaviour is therefore dependant on the extent, speed, duration and number of load cycles. With increasing duration of loading or extent of loading, the resistance against deformation increases.[2]

When tension stress is applied, a clear time-dependant development of strain can also be observed. If an applied stress leads to an immediate instantaneous strain (without time delay), which however does not lead to a permanent strain, then this is called inelastic behaviour. If a purely elastic and *inelastic* deformation component is followed by a plastic (irreversible) component, then this is called *visco-elastic deformation*.[3]

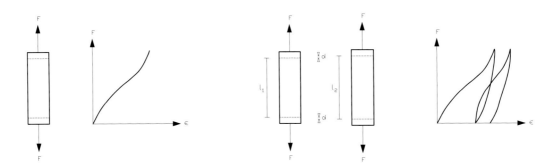

Fig. 66: *left:* Geometrical non-linearity; *middle:* Strains under repeated loading; *right:* Loading and unloading curves

1 Domininghaus, H. (1990)
2 Blum, R. (1990)
3 Föll, H. (2005)

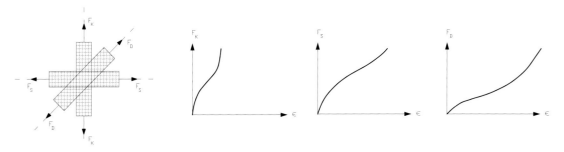

Fig. 67: Dependance of fabric behaviour on strip orientation

Anisotropic behaviour

If differing mechanical values are recorded with changing alignment of a test piece relative to a coordinate system, then these are called *anisotropic mechanical properties*. The raw fabric of woven textile sheet elements is symmetrically structured and consists of woven threads running approximately orthogonal to each other.

On account of this structuring, the properties of fabrics show a preferred direction relative to a coordinate system, and their material behaviour is described as *orthogonally anisotropic* or *orthotropic*. The dependence of the stress-strain relationship resulting from a force applied to a fabric strip on the orientation of the strip to its axis substantially influences the stiffness distribution of an equilibrium area (Fig. 67). The mechanical values for the determination of strength and stiffness must therefore be determined in many fabric directions.

Influence of the thread geometry on the strain behaviour

Woven fabric structures have, on account of the manufacturing process (see section 2.2.2.1), a characteristic waviness of the fabric threads. Weft and warp threads are curved depending on the weave type, weaving process and the coating process. If tension forces are applied to such a fabric, then displacements of the fabric structure take place as a result of the alteration of the thread curvature, which are described as *constructional strain*. As a result of this, when coated fabrics are tensioned, there are also superimposed effects from the material strain properties of the fabric threads described above and also strains from the geometrical location of the thread in the composite. In addition, material strain components in the coating interact with the thread strain.

The structural model produced by Blum represents the periodic repetition of the thread geometry in an undeformed condition. For the illustration and calculation of the fabric functions, the model is divided into unit cells, with the fabric thread being abstracted (Fig. 68).[1]

With the conventionally woven fabrics used in textile architecture, the warp thread is less curved than the weft thread, on account of the weaving manufacturing process (Fig. 69 a, b).[2] The differences of thread geometry of the warp and weft threads have the effect that less curved threads under tension loading curve less than strongly curved ones. The con-

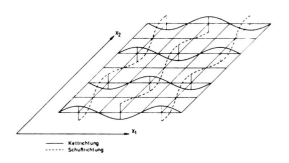 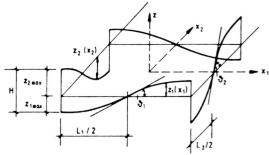

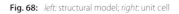

Fig. 68: *left:* structural model; *right:* unit cell

---

1 Blum, R. (1990)

2 Remark: fabrics, which are held in weft direction during weaving, form an exception here, as their thread crimp is almost equal in warp and weft direction.

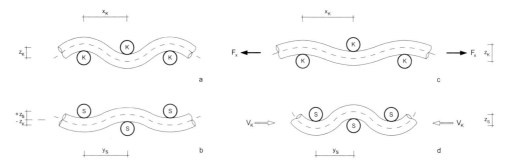

Fig. 69: Schematic layout model of the fabric threads in undeformed (left) and uniaxially tensioned *(right)* condition

siderably greater thread curvature of the weft thread is the reason for the lower strain stiffness of the fabric in the weft direction. In the axial direction of the fabric, the material has higher tensile strength than in weft direction; the stiffer warp direction therefore stretches less under loading than the weft direction.

Interaction of the fabric directions

If such an anisotropic fabric is strained in the weaker weft direction, then the weft threads stretch and the fabric elongates in the weft direction (x_K in Fig. 69 c). Because of the interaction between weft and warp threads at the crossing points and the course of the thread line, the curvature of the warp thread also alters with extension in the weft direction, and the additional curvature experienced by the warp thread causes a shortening (V_K in Fig. 69 d) of the fabric in warp direction.

If the warp direction is also under tension loading, it exerts a resistance when the weft thread is stretched. This shows that the alteration of the thread curvature in the fabric is not only dependent on the extent of the load but also from the load relationship between warp and weft. Under tension loading, any elastic body will elongate. This extension also leads to a shortening at right angles to the acting force, the ***transverse contraction*** or ***transverse strain***.

Fabrics show a considerable dependence of the strains in transverse direction on the strains in the axial direction. A tensile stress **σ**(F) applied to a fabric leads, in addition to the resulting primary strain ε_1, to contraction at right angles to the applied direction ε_2. The relationship of the axial to transverse strain is the ratio of the strain ε_1 from the primary applied load to the resulting contraction in transverse direction ε_2. It is called *Poisson's ratio μ*.

With conventional materials, the transverse contractions are related to the primary stress and the modulus of elasticity by a material constant. With fabrics, the Poisson's ratio is not constant, but dependant on the level of force applied and is usually experimentally determined through tension tests in both directions of fabric. The transverse contraction is determined by the orientation of the thread layout in the fabric and the E-moduli (Young's moduli). The different stiffness of warp and weft threads lead in this case to different contractions. The interaction between the thread directions can, for fabrics with dissimilar strain properties, even lead to the situation that the warp direction experiences negative strain under heavy loading of the weft direction. The transverse contraction is also determined by the type of edge detail of the membrane surface. In order to hold fabric in position along its edge, opposing forces have to limit or hinder the transverse strain.

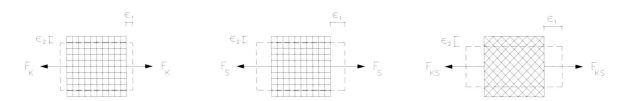

Fig. 70: Sketch of the transverse strain produced by axial strain for fabrics with dissimilar stretching properties

This can happen through flexible or stiff edge details. In both cases, the transverse contraction exerts forces transverse to the edging.

Local stress relationships between warp and weft directions determine the tension and the transverse contraction flexibilities. For structural design, it is therefore centrally important to know how strongly the load action in one fabric direction influences the forces and strains in the other. The influence of transverse contraction on the interaction of warp and weft direction is determined in biaxial tests.

The superposition of all these interactions between the two thread directions with partially reversible and partially irreversible deformation components creates a very complex behaviour of the fabric, which has to be investigated in mathematical models and requires very precise design values for the serviceability state.

2.3.1.2 Strength and stiffness of the fabric

The determination of the mechanical behaviour of coated fabrics is decisively dependant on the thread behaviour and on the geometry of the warp and weft threads in the fabric. To evaluate the strength and stiffness of a fabric, the strength and strain stiffness of the warp and weft threads in the fabric are most relevant.

To evaluate the mechanical behaviour of fabric membranes, the loadings are separated into short-term peak loads and permanent loads. The stress, which will lead to the breakage of the fabric in 11.4 years, is calculated from the long-term strength, the long-term uniaxial and biaxial loading of the material under load-dependant and load-independent actions. The stress calculated from the strength at a particular point in time within the projected lifetime with the different loading history of the material is called the residual strength (creep strength). This stress in the fabric must be less than

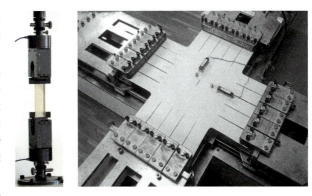

Fig. 71: *left:* Uniaxial tension test; *right:* Biaxial machine at Stuttgart University, Germany

the maximum permissible stress, which leads to breakage of the fabric. Long-term acting loads can be determined in up to 1,000 hour long laboratory tests under constant load.

The short-term strength for plastics is considerably higher than the long-term strength and thus also higher than the residual strength. The starting values for calculating the permissible bearing load of a fabric are the short-term breaking loads. The determination of the short-term strength is performed with uniaxial tests on fabric strips. The fabric strip is stretched to the breaking load under constantly increasing tension.

The short-term behaviour of fabric membranes is determined in short-term tests and gives information about the behaviour of the fabric in various loading situations. To determine the strain behaviour of fabric membranes under loading, biaxial tension tests are carried out.

In order to be able to test whether the fabric used can fulfil the structural requirements, the stress relationships in the material must be investigated under the relevant loading cases. To do this, the stress-strain relationship of the fabric

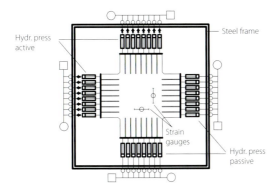
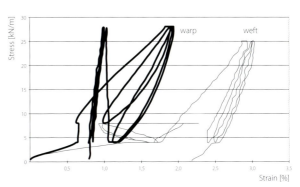

Fig. 72: *left:* Sketch of a biaxial test rig; *right:* Stress-strain diagram of a Glass/PTFE fabric

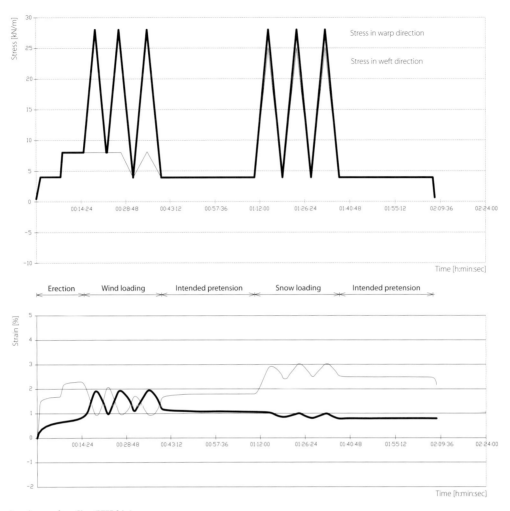

Fig. 73: Loading diagram for a Glass/PTFE fabric

is tested with different loadings in the warp and weft directions (right in Fig. 72). The forces to be applied and duration of loading were in line with the structural requirements.

Biaxial tension tests can also be performed with the expected loading history and the corresponding strains.[1] The data resulting from the tension test deliver exact details of the material behaviour under loading during a defined time frame at a defined test temperature and loading speed.

Tension test in Fig. 73:

Material: Fibertech Glass/PTFE fabric 1028/EC3, thickness: 0.9 mm, weight 1,500 g/m²

1. Loading of the warp and weft direction with the intended tension of 4 kN in ratio 1:1 (simulation of the load acting during erection)

2. Maintenance of the load (load transfer into the glass fibre fabric during erection)

3. Increase of load to 8 kN (overtension/erection)

4. Maintenance of the load at the intended stress

5. Pulling the warp direction at intervals with 28 kN (Wind loading)

6. Maintenance of the load at the intended stress

7. Simulation of snow loading with 28 kN in warp and 25 kN in weft direction

8. Reduction and maintenance of the load at the intended stress to the end of the test, reduction of load over time (creep) with reducing strains

In particular cases, the loading programme can also be produced diagonal to the warp and weft directions.

---

1 *The tension test shown in Fig 73 was performed by Labor Blum in Stuttgart (project engineers IF-Group, Reichenau).*

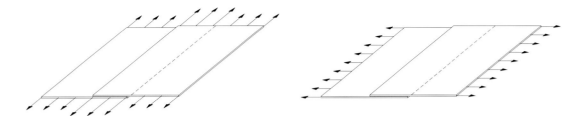

Fig. 74: Axial and transverse loading of the seam

2.3.1.3 Strength and stiffness of the seam

Curved membrane surfaces cannot be produced in any width. Manufacturing methods force the division into strips, which have to be joined to each other. Joints in wide-area structures normally have the function of applying or transferring forces. With fabric membranes, this is usually done with welded overlaps, where the forces from the fabric threads are applied through the coating into the threads of the next strip.

The strength of a membrane surface is only as high as the strength of its joints, and each joint disturbs the load-bearing behaviour of the membrane. The cause of this is, on the one hand, the material and construction of the seam, and on the other hand, that the seam is a linear load-bearing element and represents a geometrical discontinuity in the flow of forces in the curved membrane surface.

The loading of the seam depends on its location in the surface. If it lies axially, then it causes stiffening in comparison to the membrane material on account of the increased quantity of material (left in Fig. 74). If it lies transverse, then there will be a loss of strength in comparison to the fabric (right in Fig. 74).

In both cases, the seam represents a problem area, because a stress concentration can start from the folding, which appreciably reduces the lifetime of the joint. With pneumatically supported surfaces, the reduced stretching of the seam in comparison to the neighbouring material causes constrictions.

If the seam is loaded in transverse direction, then a loss of strength of approx. 20 – 25 % compared to the fabric has to be reckoned with. According to the detail and the height of the construction, the seam could slant against the axis of the seam (Fig. 75). There is often critical loading of the fabric next to the seam, which explains the fact that failures of the fabric often start from such locations if there is an overload.

The seam strength is an important parameter in the evaluation of the mechanical behaviour of a welded seam. The quality of the coating and its interaction with the fabric are the important factors for the seam strength. If force acts on the fabric transverse to the seam, then the coating is loaded in shear (Fig. 76). The strength resisting this shear loading demands a strong adhesion of the coating to the fabric. The adhesion strength of the coating on the back of the thread is therefore decisive for the strength of the seam.[1]

Fig. 75: Slanting of a welded overlap (left) and a sewn double seam (right) under transverse loading Welded seam

Fig. 76: Diagram of the interaction of coating and fabric at a welded seam

1 Bögner, H. (2004)

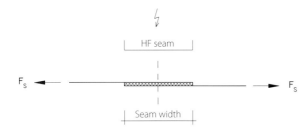
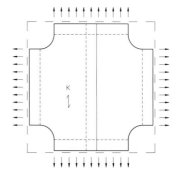

Fig. 77: Seam test/HF-welded seam

The necessary bond between raw fabric and coating is produced during production by the adhesion compound mixed with the coating mass. The higher the adhesion strength of the coating, the higher the strength of the seam.

In addition to the strength between coating and fabric, the type of welding and the welding speed are decisive for the quality of the joint. A favourable distribution of the shear stress also depends on the seam width and the elasticity behaviour of the fabric and the coating.

The adhesion strength of the coating on the fabric, the specification of the optimal welding parameters and the optimal seam width for the distribution of the shear stress can be tested in uniaxial and biaxial short-term tests (Fig. 77). The results of these tests are also temperature-dependant. With increasing temperature, the coating tends to soften and the adhesion effect and mechanical linking are reduced in their effectiveness.[1] The data from the test, produced on the basis of the seam requirements from the structural design, is used for the structural calculation.

If wider seams are used in order to improve the stiffness relationship between membrane and seam, then care must be taken that the seam can also conform to any deformations of the element, which may be necessary for a reverse cur-

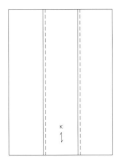 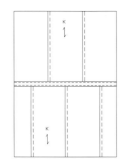 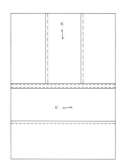 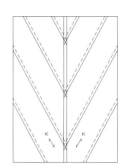

Fig. 78: Seam lines, joint arrangement

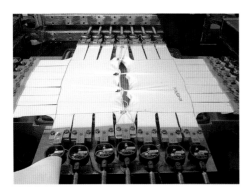  

Fig. 79: Biaxial test on a laced connection

1 Minte, J. (1981)

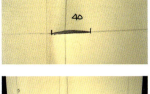

Fig. 80: *left:* Test of tear propagation for the membrane roofing of the Sony centre in Berlin, Germany; *right:* Diagram of a uniaxial te

vature. In order to avoid unclear stress conditions, the direction of the strips in the surface should be kept parallel if possible. If this cannot be organised, then different forces will act along the joints where the warp direction meets the weft direction. This should be considered in the calculation of the patterning.

During the welding process, the seam shrinks in the axial direction. The resulting welding shrinkage is taken into account in the cutting-out patterns through seam allowances. The seam is pre-tensioned during the welding process, marks are made along the seam and the strips are variously pre-stretched and lined up. After spots have been fixed, the seam is welded. Thus the strength of the seam also depends on the production conditions (see section 2.5.1.1).

Biaxial seam tests are also carried out on laced joints. The loading is increased to determine the geometry of tears and the location of tear formation. Such tests are the basis for the improvement of the interaction of eye, welded seam and lacing cord.

2.3.1.4 Tear propagation behaviour

If standing threads of a fabric are damaged by external influences under loading, then there is a danger that a tear can elongate. If the initial tears continue to lengthen in the fabric, this is called tear propagation. Most known failures of coated fabric have been caused by tear propagation, which is known to be the most common failure mode for fabrics. Stress tear concentrations through tear propagation under working and breaking loads can be investigated in biaxial tear propagation tests, which are however rarely carried out in Europe. Comparative values can be determined in uniaxial tests, but these are not suitable as a design basis.[1]

2.3.1.5 Shear stiffness

If forces, which do not act in the main direction of the fabric, act on a surface through the edge detail, then the shear stiffness of the fabric plays a considerable role. The fabric at this location must permit shear deformations between warp and weft threads. These deformations correspond to a shear distortion between warp and weft threads, which can only be

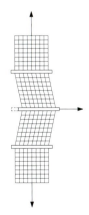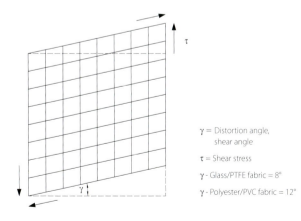

γ = Distortion angle, shear angle

τ = Shear stress

γ - Glass/PTFE fabric = 8°

γ - Polyester/PVC fabric = 12°

Fig. 81: Distortion of fabric under Shear loading

[1] Blum, R. (1990)

permitted to cause slight stress in the material. As long as the warp and weft threads do not hinder each other from sliding, they only create a slight resistance to this distortion through the coating, without excessively large stresses arising. If the displacement continues, warp and weft threads lock each other. The resistance against further displacement increases sharply; this condition is called jamming.

This circumstance also has great importance for the cutting-out of fabrics because the limiting angle, at which the resistance to the displacement increases, limits the strip width in relation to the relevant curvature.

The extent of the angle of rotation is primarily dependant on the weave type and the coating. To estimate the influence of the displacement between warp and weft threads, the relationship between the force and shear deformation is determined in tests and calculations. Different fabric materials permit different distortion angles. PTFE-coated glass fibre fabrics have a higher shear modulus compared to Polyester/PVC fabrics.

The shear strain resulting from distortion can be determined in uniaxial or biaxial shear tests. In the biaxial test arrangement, the anisotropy direction of the fabric is loaded at an angle of 45° to the main stress direction. The shear stress and the shear modulus can be calculated from the measured strains. In the uniaxial test arrangement, a fabric strip is tensioned in a test machine and held with 3 clamps. The central clamp is pulled horizontal and the displacement of the force required for the deflection is measured (Fig. 81).[1]

2.3.1.6 Kink resistance

Coated fabrics are subjected to extensive mechanical effects during erection. Above all, when folding, lifting and clamping loose panels to the primary construction, care must be taken that no localised damage occurs to the coating. Also when constructing the connections to stiff load-bearing elements, the resulting tangent angles of the forces arriving must be matched or transferred by movable components.

Breakages in the coating can expose the carrier fabric. The fabric threads are then open to environmental influences and this can lead to early failure of the construction. Glass/PTFE fabrics are, on account of their brittle glass fibres, considerably more susceptible to kink damage than PVC-coated polyester fabrics. Kinked fabric samples in a test rig are rolled over by a roller of defined weight to determine the loss of strength.

2.3.1.7 Strength-reducing effects of load, time and temperature

The load-bearing capacity of coated fabrics is reduced by long-term loading. If there is an increase of strain under constant loading over a certain period of time, then this is called load-dependant creep (relaxation) of the material. Thermoplastics show more or less pronounced creep, even at room temperature. An essential factor for creep is the time scale. Under constant stress σ, the deformation ε increases with time τ (left in Fig. 82). The longer the duration of the loading, the less is the load-bearing capacity. To investigate this, creep tension tests are carried out under constant stress and the

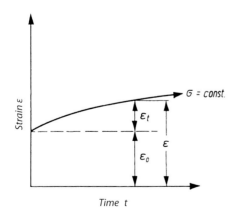

Creep curve (retardation curve)

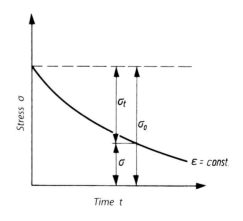

Relaxation curve

Fig. 82: Thermoplastic behaviour under loading over time, diagram of principle

---

1 Bögner, H. (2004)

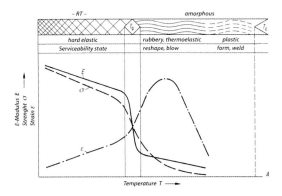
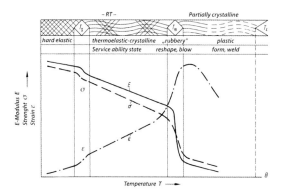

Fig. 83: Thermal condition zones for amorphous *(left)* and partially crystalline thermoplastics *(right)*

strain is recorded against time to produce creep curves. If the strain ε is kept constant under long-term loading, then the stress σ reduces in the course of time t, and there is a reduction of stress, called **relaxation** (right in Fig. 82). As a result of this, there is a delayed load reduction under constant strain.

Time-dependant deformations are also dependant on the thermal behaviour of plastics. Thermoplastics become brittle at low temperatures. With increasing temperature, they suffer, on account of their structure, a relatively large volume expansion. This results in a constant fall in elastic modulus and an increase of stiffness.

The understanding of the processes between the defined conditions of glass transition temperature T_g and the melting temperature T_m or the decomposition area T_z are the key to understanding the mechanical properties of plastics at various temperatures (Fig. 83).[1]

The temperature and load dependant creep behaviour is decisively influenced by the chosen type of material and the method of connection. For pretensioning, the erection procedure to bring about the state of pretension, this behaviour has considerable significance.

Tensioned fabrics are in an even stress condition; the stresses can displace locally in the surface. If such a fabric is erected and pretensioned, the stress distribution can, depending on the fabric material used, last several weeks under some circumstances on account of the creep behaviour. This behaviour is especially pronounced with the relatively stiff Glass/PTFE membranes. The installation of equipment for subsequent retensioning can make the erection easier if creep is expected.

PTFE-coated glass fabrics are considerably more prone to creep at low temperatures than polyester/PVC fabrics. There is a crystal change in the material PTFE at about 19 °C leading to a shortening of the molecule chains. This shortening reinforces the stiffening of the fabric at low temperatures through the manufactur still more, which has an effect on the processing temperature of Glass-PTFE fabrics during erection.

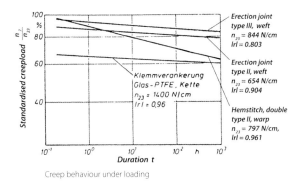
Creep behaviour under loading

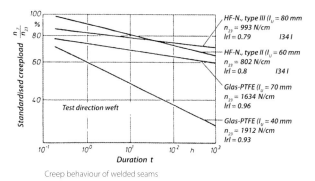
Creep behaviour of welded seams

Fig. 84: Creep behaviour of membrane joints at room temperature according to Minte

1 Dominighaus, H. (1992)

So erection at low temperatures is only possible with considerably more effort than at higher temperatures.

According to measurements at the GH-Essen University, approximately twice the force is necessary to tension a Glass/PTFE fabric at low temperatures in order to achieve the required strain values. If erected with insufficient force potential, the tension could fall off after strong wind. This could lead to fluttering of the fabric, which causes a high dynamic loading and could destroy the construction. It is therefore better to monitor and control the tensioning movement in addition to the tensioning force. If no measured temperature values are available, it is also advisable to retension the membrane at a later date. Thus the temperature is also a relevant parameter for the tensioning process.[1] Because of the time and temperature-dependant effects on the mechanical tensioning of fabrics, the estimation of the time schedule for performing the tensioning is therefore of central importance.

Because the applied tension reduces at constant strain during the lifetime on account of relaxation, the residual strains have to be determined in tests and the geometrical dimensions of the membrane corrected (see section 2.4.2). To determine the different compensation values in warp and weft directions for the patterning of the strips, data regarding creep strain under the applied tension and the reduction of tension over the lifetime is also important. To determine the creep and relaxation behaviour, long-term tests are necessary.

One further significant factor for estimating the load-bearing behaviour of a membrane surface with time is the method of connection. To investigate this, surface and edge connections are loaded with standardised creep loads and their creep behaviour evaluated against creep curves.[2] Because of the complex material behaviour under loading and with time, it makes a considerable difference for the design whether the structure is to be designed for a few years life or for many decades.

For the dimensioning and determination of the load-bearing capability of textile fabric membranes and their connections, a system of reduction factors is used, derived from breaking loads determined in tests. This takes into consideration the strength-reducing influence of loading (A0), time (A1), UV radiation (A2), temperature (A3) and quality variation in production (A4).[3]

---

1 Saxe, K.; Kürten, R. (1992)

2 Minte, J. (1981)

3 Sobek, W.; Speth, M. (1995)

2.4 Fabrication of coated fabrics

In order to be able to use fabric membranes and foils as load bearing elements, they have to be divided into strips, cut and connected together according to geometrical and structural design requirements. The preparation of strips of fabric from the roll material and the joining of the pieces, either by bonding or mechanically, is described as *fabrication*.

The influential factors for cutting out the pieces resulting from the form, the bearing behaviour and the erection interact with this process and have a great influence on the practical side of membrane construction.

2.4.1 Development

The equilibrium shape of the curved membrane surface, which has been calculated by the structural designer and experimentally tested, must be divided into two-dimensional strips (projected onto a plane), which can be cut from the roll material delivered from the weaving mill up to about 5 m wide. This process of creating two-dimensional cutting patterns from the three-dimensional form is called development. The development of the individual pieces in two dimensions fixes the layout of the strips from the roll and thus the anisotropy axes for the distribution of stiffness in the membrane.

Geometrical surfaces can only be mathematically developed into two-dimensional surfaces when their Gaussian curvature at every point is zero. Curved structures, which have found their own form, always have a negative Gaussian curvature and so cannot be developed. So to produce the two-dimensional pieces, the different lengths from the roll resulting from the mathematical-geometrical constraints have to be compensated through a patterning calculation. This calculation is based on material properties, which can predict the behaviour of the material under deformation.

The shape has to be formed with no folds in the material, or very few. Because the forming of the shape is dependant on the material, there are two very different ways of projecting onto a plane: true-length projection for woven sheets and true-angle projection for foils (see Fig. 85).

For fabrics, the application of a shear deformation, which displaces one edge parallel to the opposing edge, allows the surface to be projected onto a plane with correct lengths. This parallel deformation should cause no, or very slight, stresses in the material. This type of deformation only succeeds with materials of low shear stiffness. Woven fabrics have a low shear stiffness and are therefore suitable for a change of angle between warp and weft. As long as warp and weft threads do not hinder each other when deformed in this way, they only offer slight resistance to shear distortion, without too large stresses developing.[1]

The extent of the angle of rotation is mainly dependant on the weave type and the coating. Tests and calculations are performed to determine the relationship between force and shear deformation and evaluate the influence of the deformation between warp and weft threads. Different fabric materials permit different angles of deformation. The shear modulus as a physical value is of the greatest importance for the production of fabrics (see section 2.3.1.5).

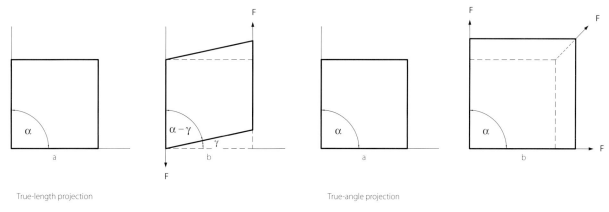

True-length projection

True-angle projection

Fig. 85: Methods of projection

---

1 Blum, R. (1990)

Fig. 86: Membrane strips compensated in warp and weft direction before anchoring to fixed points

For isotropic foils, the true-angle projection causes a non-homogeneous strain in the material, with every point on the surface having a different state of stress. For thin foils, this state of stress is below the plastic range, for thick foils such high stresses are created that the material flows.[1] When using foils as a pneumatic envelope, this strain is brought about by the pressure differences.

In both projection methods, it can be seen that the distortion can be first evaluated and the projection onto a plane is only possible through the utilisation of the material properties.

The stretchability of the material used has a direct relation with the possible curvature of a membrane construction. Because various materials also permit various angles of distortion, investigations must be undertaken to establish which element deformations in the material result from which curvature. Stiff fabrics, which only permit slight rhomboidal deformation, require very precise pattern calculation. The greater deformability of flexible materials helps to relieve peaks of stress, which mostly leads to smaller deformations in the overall system.

In addition to the requirement to keep the cutting out waste as low as possible, the width of the material from the roll plays an essential role in the cutting-out calculation. So this calculation is generally more problematic for smaller structures than for larger. The proportion of strips with maximum width is higher in relation to the overall dimension for smaller surfaces than with larger structures, where the length distortion for projecting the three-dimensional cut-out is relatively low for each individual strip.[2]

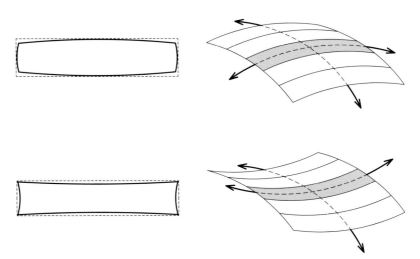

Fig. 87: Curved strips for synclastic *(left)* and anticlastic *(right)* surfaces

1 Blum, R. (2004)

2 Moncrieff, E.; Gründig, L. (1999)

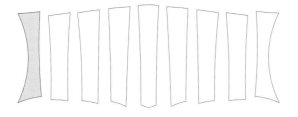 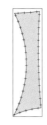 

Fig. 88: Layout of strips, cut-out strips, intended geometry

2.4.2 Compensation, strip layout

Because flexible materials deform under the influence of loading (constructional strain) and temperature (creep, relaxation), the resulting strains have to be be determined in tests and the geometrical dimensions of the material have to be corrected so that the structural requirements are still fulfilled at the end of the lifetime. The correction of the geometrically developed surface for the stretching resulting from the structural factors is described as *compensation*. If there are negative strains in certain areas as a result of the calculated stress distribution, then material must be added. This is then called *decompensation*. For membrane surfaces, this is often necessary in areas with increased stiffness near fixings.

Loads applied to fabrics with dissimilar stretch properties in warp and weft directions also cause different deformations in the main anistropy direction. In this case, the stiffer warp direction is mostly considerably less compensated than the weft. To determine the compensation values, which are different for the warp and weft directions, data for creep strain under the applied tension and the reduction of stress over the lifetime is important.

The load transfer in lightweight structures can be through surfaces curved in the same direction or in opposite directions. The type of curvature also determines the cutting out of the individual strips. In order to be able to make membrane surfaces curved in the same direction or in opposing directions out of individual fabric strips, the strips are cut with a curve along the edges.

Surfaces with positive Gaussian curvature receive strips with edges cut convex, and surfaces with negative Gaussian curvature receive edges cut concave (Fig. 87).

In order to end up with the correct geometry of a double-curved surface, the membrane must be cut in strips along prescribed seam lines (normally geodetic lines). For this reason, the strip widths are to be coordinated with the curvature, the structural requirements and the available roll widths. Connecting together the flat strips corrected by the compensation values produces the intended geometry.

2.4.3 Criteria for the patterning

The establishment of the three-dimensional cutting patterns with the determination of the arrangement and dimensioning of the individual membrane strips on the membrane surface is called patterning. It is primarily dependant on the shape of the surface, which is to be produced.

Model-based processes of patterning with empirically estimated compensation for the increase of strain in the tensioned state are only suitable for the preliminary design stage. In the further course of the project, the patterning calculation has to be performed analytically. The calculated compensation values are anticipated on the basis of biaxial tests under long-term loading of the actual stretch in the tensioned state through reduction in size of the individual pieces.[1]

One important parameter for the determination of the layout of the strips, in addition to the optical impression of the membrane structure, is the best possible exploitation of the available material strengths – observing the maximum permissible deformations and buildability. The calculation of cutting patterns also has a decisive effect on the economy of a membrane structure. It should therefore be an integral part of the form finding.

The purpose of the patterning calculation is to reproduce the calculated equilibrium shape as precisely as possible considering the aspects stated above.

In summary, the criteria for the patterning calculation can be divided into: optical criteria, topological criteria, practical erection criteria and processing criteria (Fig. 89).

---

1 Mühlberger, H. (1984)

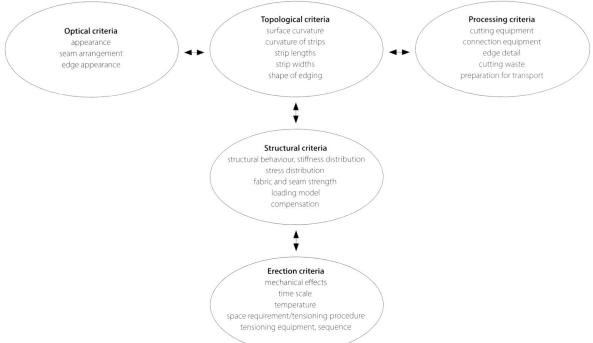

Fig. 89: Criteria for the calculation of patterning

2.4.3.1 Topological and structural criteria

Stiffness distribution – layout of the strips

Because of the anisotropic behaviour of fabrics, the stiffness distribution in the membrane is fixed with the determination of the layout of the strips (strip arrangement). The cutting direction thus plays a considerable role in the direction of the load transfer in the membrane surface and has an influence on the structural system up to the deformation of the primary structure.

The calculation of the patterning depends primarily on the shape of the surface to be created. Before the fabric can be cut to size in strips, the main anistropy direction of the fabric relative to the main direction of curvature must be determined on the basis of calculations.

The stresses, which arise in the membrane surface and its edges, and the resulting strains in the fabric determine the choice of strip layout. If the membrane surface, which is to be created, corresponds to the smallest surface, in which the stresses are equal in all directions, and fabric is used with the same strain in both thread directions, then the layout of the strips can be chosen at will. The material is then utilised optimally and local overstress is mostly avoided.

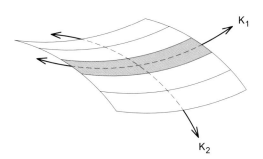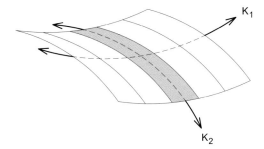

Fig. 90: Stiffness distribution / layout of the strips in main direction of curvature

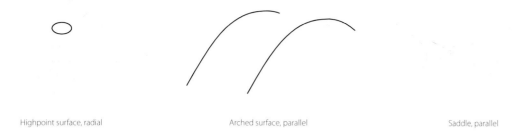

Highpoint surface, radialArched surface, parallelSaddle, parallel

Fig. 91: Strip layouts with warp in the main bearing direction

In the practice, this is actually very seldom the case; membrane surfaces normally consist of materials, whose main stresses vary from each other.

In such shapes, the strips are mostly laid in the main loading direction, in order to be able to optimally level the surface stresses arising through compensation of the main anisotropy direction. The main curvatures must in this case be chosen so that the membrane stresses caused by external loads can be borne by the type of fabric used.

The main bearing direction of a membrane surface is the direction of the greatest strain. Because the warp thread in woven fabrics has a higher E modulus than the weft thread, with the result that the material has a higher tensile strength in the long direction of the fabric than in the weft direction, care is taken in the determination of the strip layout that the warp is mostly laid in the main bearing direction, because less deflection can then be expected under loading. Another advantage of this strip layout is that the weaker seams do not have to bear the highest possible stress in the transverse direction.

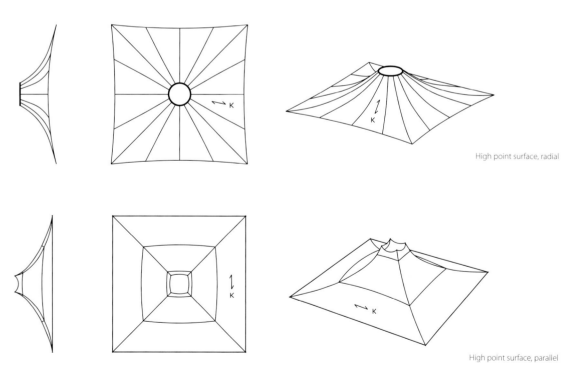

High point surface, radial

High point surface, parallel

Fig. 92: Possible strip layouts for highpoint membranes

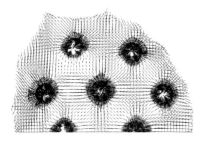 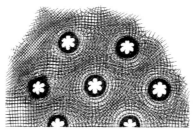 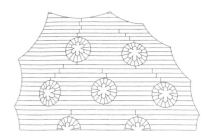

Fig. 93: Main stresses and strip layout for a swimming pool roofing in Malaysia

The arrangement of the strips in the main bearing direction can be parallel, radial or a combination of these two cutting pattern types. Construction with the strips arranged parallel normally allows a good stress distribution to be reached.

Radial patterns use more material than parallel layouts on account of the higher cutting waste and are considerably harder to dimension and compensate.

Because the deformation is reduced by the increasing stiffness of a construction, the material is also an important factor in the determination of the strip layout. Membranes with low stretch accordingly effect an increase of stiffness. The aim is to choose the layout of the strips so that the cutting out brings the least distortion.

The determination of the main anisotropy direction is subject to special conditions for surfaces supported at a high point. The constant reduction of area towards the high point causes stress concentrations and high radial forces at this location. This results in higher strength requirements with less material capacity near the high point.

With high point structures, it is possible to arrange the strips either parallel or radial to the high point. If the stiffer warp direction is in the main bearing direction (top in Fig. 92), then less deflection of the structure can indeed be expected, but with stronger materials the weft direction must be considerably more strongly compensated in order to achieve a homogeneous stress distribution.[1] If the warp is laid in the direction of the ring forces (bottom in Fig. 92), then force can certainly be brought into the ring in the stiff direction, but higher cutting wastage can make this solution uneconomical.

For structural and production reasons, the parallel layout of strips is only practical for high point structures with straight, stiff edging and an angular high point arrangement.

With linked high point structures, surfaces with strong curvature meet surfaces with weak curvature. Here it is sensible on account of the geometry to arrange a combination of parallel and radial strips (Fig. 93). The stress distribution under various loading conditions is decisive in this case.

For a 9,000 m² swimming pool roof in Kuala Lumpur (Malaysia), the force concentration acting on the high point from the loading case "snow" can be seen clearly; there is a constant increase of stress towards the centre point (left in Fig. 93). Under the loading case "wind uplift", the forces run in waves around the high points (middle in Fig. 93). The analysis of the stress distribution under the decisive loading cases produced the optimal strip arrangement with combined parallel and radial strips (right in Fig. 93). The stress concentrations under loads acting downwards determine the radially arranged membrane strips. In order to make the passing of forces as undisturbed as possible, the arrangement has as few transverse seams or transverse strips as possible.[2]

Surface curvature – layout of strips – strip shape

Because the trimmed strips are aligned with the main curvatures, the patterning has a major influence on the curvature properties of a membrane surface. The angle errors resulting from the non-developable shape have to be compensated in the patterning so as to optimise the uniform stress distribution in the surface and to be matched to the available material capacity. This is mostly achieved when the main stress directions are in agreement with the main curvature directions.

The extent of the curvature of the surface is directly related to the way of cutting out the individual strips and the strip division for stiffness distribution. Strongly curved surfaces do indeed have a positive influence on the load-bearing behaviour, but make the patterning more difficult. When the surface is strongly curved and the material is stiff, precise patterning calculation is necessary to achieve a homogeneous distribution of the stress in the membrane surface. The strip layout plays an essential role. So it is possible rather for

---

1 *Moncrief, E.; Gründig, L.; Ströbel, D. (1999)*

2 *Schlaich, J.; Bergermann, R.; Göppert, K. (1999)*

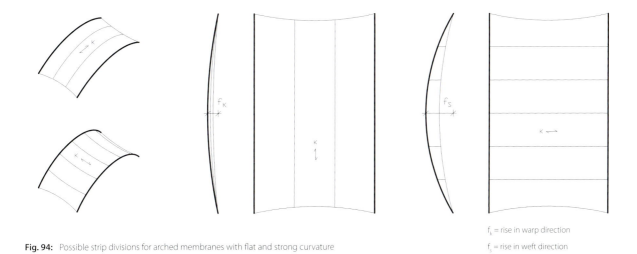

Fig. 94: Possible strip divisions for arched membranes with flat and strong curvature

f_k = rise in warp direction
f_s = rise in weft direction

less curved surfaces (f_k in Fig. 94), to turn the warp and weft directions when cutting out than with larger arch rises (f_s in Fig. 94), because smaller curvatures produce smaller angle errors from fabric distortion.[1]

The more strongly a surface is curved and the longer the strips cut out, so the greater are the distortions, which have to be compensated by calculation.[2] A more favourable distribution of stress can be achieved for strongly curved surfaces through the arrangement of shorter and narrower strips. The relationship of the strip width to the strip length thus also plays a considerable role in load transfer.[3]

When deciding the strip layout for less curved surfaces, attention should be paid to possible changes of load transfer direction during high winds. In this case, the stresses in the material dissipate in the other thread direction, where they lead to a shortening of the thread. The fabric becomes weaker till it turns inside out. There is a change of shape from an anticlastic to a synclastic surface. Wind uplift ropes fixed to the inside or the outside of the membrane surface or loosely on the surface can hinder this. The wind uplift rope can be fabricated into the membrane surface or tensioned next to the surface. Such a measure can also be sensible for more strongly curved surfaces where long-term uniform wind loading is to be expected (Fig. 95).

Because the length differences from the compensation can only be absorbed in the corners of the cut-out strips, the

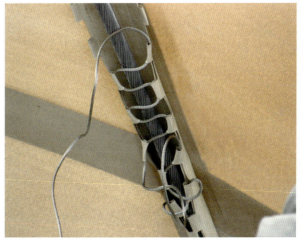

Fig. 95: Diagonally arranged uplift ropes underneath an awning surface

---

1   Essrich, R. (2004)

2   Ziegler R.; Wagner, W. (2001)

3   Remark: No publication is known to the author, in which the relationship of arch rises and main curvatures to possible spans and strip layouts of anistropic membrane surfaces has been investigated. Generally valid statements about the ideal strip layout can therefore not be made at this point

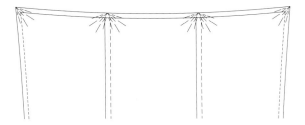 

Fig. 96: Load singularities depending on the width of cut pieces

material in this area must permit potential large distortions between warp and weft threads though angular twisting. With larger strip widths, load singularities can be expected in acutely angled corner areas because of insufficient material capacity. Reinforcement measures must be provided at such locations.

In order to avoid singular overloading in the corner areas, it can be sensible to reduce the cut-out size of the individual strips. The arrangement of double or triple strips can considerably reduce the appearance of undesired stresses in the corner areas. The calculated overall compensation value can in this way be distributed over many corners of the joined up strip.

The situation that the cutting out of smaller membrane surfaces is generally harder to design than larger surfaces follows from the fact that a large membrane surface generally consists of a multitude of cut-out strips. The distortions arising during patterning are thus relatively small for each individual strip.[1]

The resulting arrangement with many seams is not, however, ideal for tensioning process. Unfavourable stiffness conditions between seam and fabric material make the tensioning much harder.

Shape of the edging

Like the surface curvature, the geometry of the edging of a membrane surface also has an influence on the load-bearing and deformation behaviour. Load-bearing elements made of membrane material can be edged with flexible or stiff edging (see section 2.6).

The rule for mechanically pretensioned membranes with curved edging of panels is that the curvature of the edge line must be concave. An alteration of the curvature radius of the edge line also produces an alteration of the curvature properties of the membrane surface. For curved, flexible edges, the parameters, which influence the geometry of the surface, are the edge rope tension force (S_S), the edge rope radius (R) and the membrane stress resulting from the load on the material (S_M).[2]

The relationship between edge rope force and the membrane stress determines the geometry of the edge; the shape of the membrane is thus a function of the edge geometry.[3]

The more pronounced the curvature of the edge rope radius is, the higher will be the maximum and the smaller the minimum Gaussian curvature. The increase of the edge curvature radius is accompanied by an increase of stress in the material. Ideal conditions are denoted by a minimal area, in which the extent of the minimum and maximum curvature is identical.

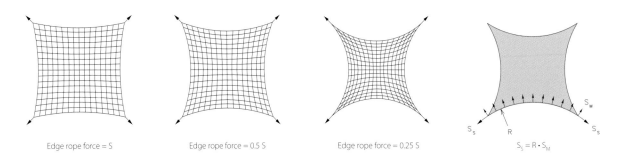

Edge rope force = S   Edge rope force = 0.5 S   Edge rope force = 0.25 S $S_S = R \cdot S_M$

Fig. 97: Alteration of the main direction of curvature of a four-point awning through variation of the edge rope radius and edge rope force

1 Moncrief, E.; Gründig, L.; Ströbel, D. (1999)
2 Ziegler, R.; Wagner, W. (2001)
3 Ziegler, R. (2001)

2.4.3.2 Erection criteria

In addition to the determination of the strip layout favourable for the deformation behaviour according to structural and topological criteria and production limitations like strip width, cutting waste and seamed joints, the division of a membrane surface into strips is also subject to limitations regarding erection. It becomes particularly clear during the tensioning process, how important the material used, its cutting patterns, the shape of the surface and the type of edging are for the erection. This particularly makes the tensioning scheme, which can have a considerable effect on the overall project costs, economically significant in the erection process.

Tensioning travel, tensioning direction – cut-out pattern direction

In order to make a membrane sufficiently load-bearing, it has to be biaxially tensioned in the relevant curve and fixed. The calculated share of compensation must be pulled out of the material through the edging and its corners.

The cost of implementing a tensioning procedure is determined mainly by the arrangement of the individual panels. Apart from the optical appearance of the seams and the structural conditions in a membrane surface, all erection measures, like the dimensioning and installation of the tensioning equipment, measures to stabilise the primary construction and the arrangement of scaffolding, depend on the orientation of the strips.

Because varying forces have to be applied according to the compensation and the length of tensioning travel in order to pull a membrane into position, the cutting pattern direction plays an essential role for the tensioning process. The calculated shortening of the considerably more compensated weft direction is mostly pulled out of the membrane surface out of the long or transverse direction (Fig. 98). To determine the primary tensioning direction, it should be considered how large the force necessary to pull the tensioning travel is and how much the associated erection cost is.

Under the effect of loading, the main anisotropy directions shorten against one another. For the tensioning process, it is thus of central importance how load acting in one fabric direction influences the forces and strains in the other. It will be necessary to reconsider the layout of the strips for surface forms, in which no interaction between the warp (K) and weft directions can take place during the tensioning process.

The use of fabrics with dissimilar stretch properties can be very advantageous for the erection of geometries with different lengths in warp and weft directions. When using such fabrics, the ideal case is to arrange the larger part of the compensation in the weaker weft direction. The strain in the stiffer warp is very slight, and is therefore little compensated. If possible, the strip layout will be arranged so that after testing all the criteria stated above, the desired biaxial pretension will be produced solely by tensioning in the weft direction. In order to build up enough stress in the warp direction by

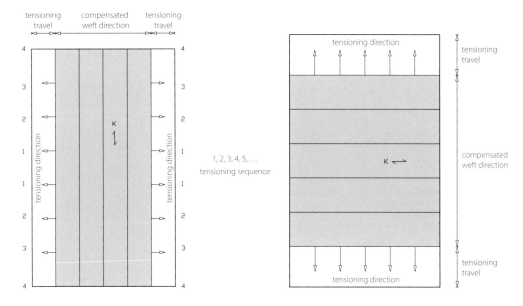

Fig. 98: Diagram of idealised tensioning procedure with limitation of transverse strain

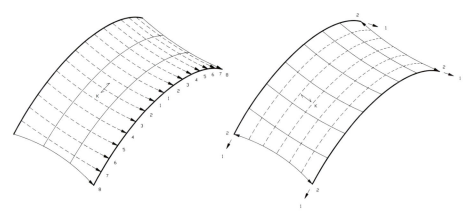

Fig. 99: Tensioning direction and sequence of arched membranes depending on the direction of cutting out

general tensioning in the weft direction, edge deformation perpendicular to the tensioning direction must be prevented. This method of tensioning can be advantageous for erection because working time and pretensioning equipment can be reduced.

If the surface has to be tensioned between two arches, as is often the case with sports stadium roofs, the membrane can be unrolled parallel or perpendicular to an arch-shaped truss. Depending on the material, arch curvature, edge geometry and edge detail, the direction of the strips makes a large difference for the tensioning process and the related measures. According to information from erection companies, there is a considerable scope for saving erection costs here, considering the space taken up by erection equipment.

Fig. 100: Pulling in and tensioning the membrane for the stand roofing at stadium Volkswagen Arena Wolfsburg, Germany

Fig. 101: Pulling in and tensioning the membrane for the stand roofing at Estádio Intermunicipal Faro, Portugal

If a membrane surface unrolled parallel to the arch truss is compared with one unrolled perpendicular to it, then it can be noticed that when tensioning the strips perpendicular to the arch truss (left in Fig. 99), the pretension has to be applied at more points than with the variant (right in Fig. 99). This requires more work and equipment.

Apart from setting up and operating the erection equipment at the location where the force is applied, this extra work is often caused by the type of edge detail.

If the use of single clamping plate edge elements cannot be avoided because of structural requirements, then the forces must be applied starting from the centre and the edge elements installed singly, adjusted and fixed (Fig. 100).

If the warp direction is perpendicular to the arch truss and the surface has the right curvature and has a keder rail edge detail, then the surface can be pulled through the rail and tensioned quickly and easily (Fig. 101).

When tensioning membrane surfaces spanned between two stiff edges, care must be taken with the loading on the seams.

If the tensioning is in warp direction, then the doubling of stiffness along the seams and the welding shrinkage must be taken into consideration. If the direction of cutting out is transverse to the tensioning direction, then suitable measures must be considered to avoid overstressing the transverse seam.

Large distortions are to be expected in the fabric of surfaces with pronounced curvature; the strips must be considerably more compensated then less curved surfaces. This is particularly clear for high point surfaces with heavy fabric. The choice can be made here to arrange the division of strips either with the warp direction of the strips radial around the high point or parallel to the lower edging.

If the stiffer warp lies in the main load-bearing direction, then a much higher compensation of the weft direction is required in order to achieve a uniform stress distribution (left in Fig. 103). A higher force is required there when tensioning to the edges than with parallel strips (right in Fig. 103).[1] Where the strips run parallel to the edging, care should be taken not to overload the seams.

If the tensioning is exclusively by vertical jacking of the high point, then care must be taken that the permissible membrane stress in this sensitive area is not exceeded during tensioning. If the tension is peripheral to the edge, less force is required and the stresses can distribute better in the fabric. The expense of installing stretching equipment along the edges or at the corners can however be expensive.

To determine the primary tensioning direction, it must be considered how large the force required for tensioning is. It is mostly better to pull a membrane surface over a long travel with low force than to pull a short travel with high force. The decision, as to from which direction the shortening is simpler and thus cheaper to pull, thus depends on the strip layout. When pulling over longer travel distances, the strains should be led away from the centre as uniformly as possible. This takes the interaction between warp and weft threads in-

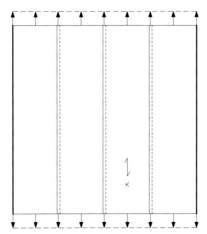
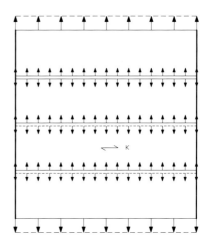

Fig. 102: Seam loading during tensioning

1 Moncrief, E.; Gründig, L.; Ströbel, D. (1999)

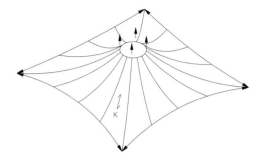

Fig. 103: Tensioning of high point surfaces with different strip layouts

to account; sideways straying of the forces because of force deflection is made less likely.[1]

In cases of topology where the membrane material has to be tensioned over equally long travel distances with the same level of force, the use of fabrics with nearly equal stretching properties in warp and weft directions (Precontraint fabric from Ferrari) can be advantageous for erection. It should, however, be remarked that the force required to span such fabrics is generally higher, making their use less favourable for relatively flat membrane surfaces with low fabric distortion.

If strongly curved surfaces span over large distances, then heavily compensated and long strip lengths are needed. Special care must be taken that sufficient material is available in all areas of the patterned pieces.

Space required and access for carrying out the tensioning

An important practical criterion for the determination of the cutting patterns is the space required during erection for the tensioning equipment, temporary construction and scaffolding. The determination of the strip lengths and and thus the primary tensioning direction also depends, for reasons of practicality, on the types of fixings available for the tensioning devices and temporary construction to be used. In this case, care should be taken with the dimensioning of the erection equipment. A large number of tensioning devices and temporary constructions can take a long time to assemble and cause more working time and high eventual costs.

The strip layout determined as favourable for the deformation behaviour is not always optimal for erection. It can be that the direction of the cut-out strips has to be altered because otherwise the pieces can only be erected with difficulty or not at all. Therefore a workable erection scheme should be produced at the design stage, taking into consideration the provision of sufficient fixing locations for tensioning equipment and their anchorage while tensioning. This shows that the determination of the strip layout also has to consider practical considerations for the tensioning process.

Fig. 104: Tensioning of high point surfaces with different strip layouts

1 Essrich, R. (2004)

2.4.4 Cutting out the pieces

Cutting out denotes the division of one or more textile surfaces (layers) according to dimensions or cutting out from pattern drawings as part of fabrication.[1]

The patterning and design are normally carried out by the engineer. Surface areas, edge details and seams are determined in discussion with the architect. The cutting out of the pieces and the joining, packaging and delivery are undertaken by the fabricator.

The purpose of cutting out is to reproduce the calculated patterning as precisely as possible. This is done by translating the cutting drawings of the individual strips onto the fabric from the roll and cutting the pieces out. The cutting drawings contain the essential specification of the material, the details of the joints and edges and the cutting shapes of the strips. Seam widths and seam allowances are given to enable the strips to be joined, and also length tolerances and production lengths.

The material information contains areas, maximum strip widths and type of fabric. The fabric direction and tearing strength in warp and weft direction are also given.

The type of edge and corner details and the joints in the surface are given in the drawings. There is also a reference to all related detail drawings (a in Fig. 105). Information about the geometry of the pieces to cut out can be read from the drawings of the individual pieces. These also include all coordinates and lengths of the compensated edges. The drawing of the piece is projected by the fabricator onto the fabric material with additional details about cutting waste and the cutting window of the machine (b in Fig. 105). The system lines denote the seam, webbing and rope axes (c in Fig. 105).

To join the pieces, the edges of the strips are marked so that no undesirable folds can occur during seaming, welding or glueing (Fig. 106).

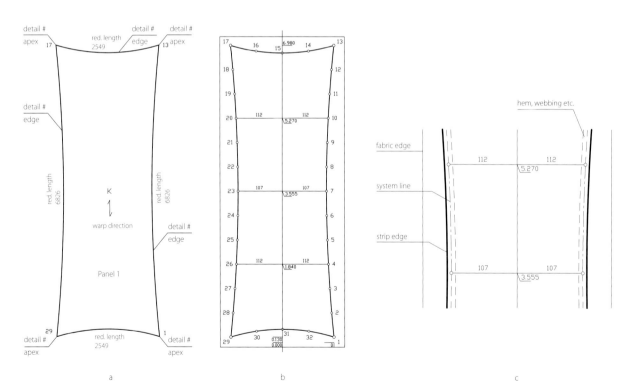

Fig. 105: Schematic cutting-out diagram of an anticlastic membrane surface

---

1 Burkhard, W. (1998)

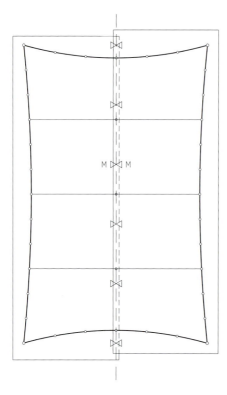

Fig. 106: Marking for the edges of the strip

The cutting out of coated fabrics is either done with hand tools or a cutting machine according to cutting capacity, layer thickness and edge geometry.

Most cutting is done on a machine with defined cutters. Depending of radius, strip length and layer thickness, these can be pull, push, round or oscillating cutters in the automatic cutting process or manual cutting of the layers using hand tools.

Thermic cutting processes are not suitable on account of the risk of the cut edges sticking.[1] The use of laser cutters is also problematic because of the chlorine-air mixture discharged during the cutting process. Ultrasound cutting machines can in principle be used, in combination with a vacuum table, but are not normally economical because of the cost. The use of water jet cutting for the cutting out of fabrics is unknown.

When there are a a number of strips to be cut with the same geometry, pre-cut templates made of thicker material can be used. The strips can be cut out singly or in piles according to the geometry and other requirements. The finished dimensions (cut lengths) must be checked in any case.

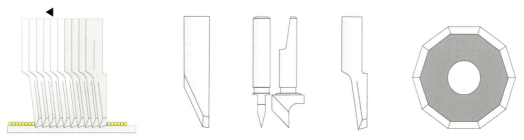

Fig. 107: Cutter tip shapes

Fig. 108: *left, centre:* Mechanical cutting machine with round cutter; *right:* Manual vertical knife machine

1 Steckelbach, C. (2005)

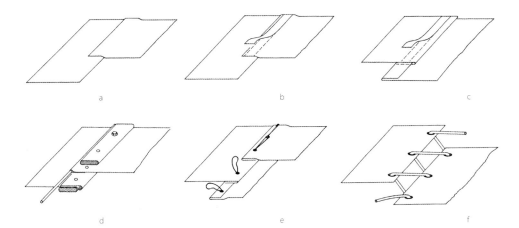

a overlapping welded seam
b sewn seam
c lapped sewn seam
d clamping plate joint
e looped joint
f laced joint

Fig. 109: Surface joint types

2.5 Methods of jointing surfaces

In order to obtain a load-bearing surface, the fabric cut into strips is joined to form panels. Permanent surface joints are made by the fabricator, and temporary or reusable joints are normally carried out on the construction site. The most important joints are:

- Permanent joints – welded seams, combination seams, sewn seams, glued seams

- Temporary or reusable joints – clamping plates and keder rail joints, looped and laced joints

2.5.1 Permanent surface joints

The material cut out from the roll is joined by the fabricator to form load-bearing panels. Various types of permanent joints are used for fabric strips according to material, structural and building specification. These can be welded, sewn, welded and sewn, or glued joints. To ensure load transfer, joints in the fabric must be able to transfer the forces out of the load-bearing threads in the fabric and transfer them to the threads of the conjoined fabric through mechanical joining or bonding. Permanent joints are normally flexible. There are, however, differences of stiffness between the two fabric strips to be joined and the seam. Obstructions to deformation along the seam joint produce irregularities in the overall shape of the membrane. Important factors to observe in the specification of such joints are the different thread locations following from the anisotropic behaviour of the fabric, the adhesion strength of the fabric coating, the seam widths and the seaming or welding process.

2.5.1.1 Welded seams

The welded seam is the most commonly used means of jointing in membrane construction. Welded seams in membrane construction can be produced as overlapping joints with varying overlaps or as butt joints with cover strip. In addition to joints in the membrane surface, welded seams may also be edge sleeve details, reinforcements, keder or webbing seams. Welded joints are normally produced at the works, where partial sheets of up to approx. 5,000 m² can be produced from roll material 5 m wide.

The strength of welded joints depends on welding process and processing temperature and is about 60 – 95 % of the fabric strength. The force transfer is through shear loading of the coating. The quality of the coating and its adhesion to the fabric produce an effect similar to form-fit and determine the strength of the joint. Welded seams produce an abrupt increase in the stiffness at the seam. Welded seams are watertight and normally UV-stable.

The welding of coated fabrics is only possible with thermoplastics. Plastics are poor conductors of heat, so the welding of thin fabrics makes fewer problems than the welding of thick layers. The welding process joins together two or more pieces of fabric of the same material under the influence of heat without the use of additional adhesive. The process can be with or without pressure. The two most commonly used processes for the jointing of coated fabric pieces are high frequency welding and hot element welding. While hot element welding thermally softens the surfaces of the pieces to be joined and presses them together with a defined pressure, high-frequency welding forms a largely homogeneous

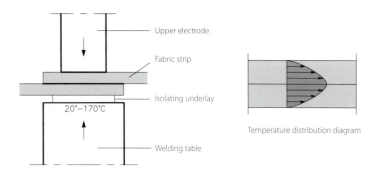 

Fig. 110: *left; centre:* Diagram of HF welding; *right:* HF welding plant

joint in which the entire coating thickness of the two material sides to be joined are integrated.

High frequency welding

The cut-out fabric strips (single templates) are joined together in this welding process through pressure and heat. Only thermoplastics with polar molecule arrangement can be welded using this process. If such thermoplastics are placed under a field of high-frequency radiation, certain groups of molecules oscillate according to the frequency, which leads to a warming of the material. The thermoplastic is warmed between a cold and a temperature-controlled electrode, brought to viscose flowing and pressed together (see Fig. 110). After the electricity is switched off, the seam has to cool under pressure in order that the melted mass can harden and the restoring force in the material can no longer become effective.[1]

Because not only the coating but also the fabric are heated, the quality of the joint is dependant on the pressure, the shape of the press, the processing temperature and the welding speed. The average weld seam width is between 50 – 80 mm. A wider seam can increase the strength of the weld for heavier fabrics. When making such wide seams, there can be a problem with the coating "swimming out" from the welding area. This can be avoided through the use of a knurled welding electrode (b in Fig. 111). Full-surface welding in the entire area of the seam can be achieved with a flat electrode (a in Fig. 111). Special electrodes can be used to make intentional breaks in the surface of the seam, which makes the seams more controllable.[2]

PVC-coated polyester fabrics and aramide fabrics welded with the high-frequency process can achieve a strength of about 90 % of the fabric strength at room temperature and approx. 60 % at 70 °C.

Surface paints containing fluorine and laminated foils prevent because of their high melting point a homogeneous connection and have to be removed where the joint is to be

a Flat weld seam   b Knurled weld seam   c Welding bead   c Seam junction

Fig. 111: HF seam types

---

1 Holtermann, U. (2004)

2 Rudorf-Witrin, W. (2004)

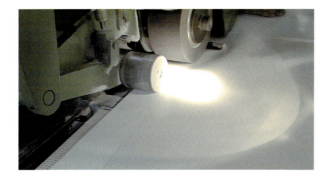

Fig. 112: Grinding off the PVDF paint

made (Fig. 112). Recently available fabrics with modified fluorine paints can be welded together without grinding off the paint, just like fabrics with acrylic paints.

To establish the optimal welding parameters for highly loaded seams, test pieces are prepared in warp and weft direction (Fig. 113). It is especially helpful to carry out such tests on welded-on erection lugs. Control tests are made before every start of work, at the change of shift and when changing the material. If the fabric is welded, then the seam is tested using a scrape test. All the parameters and testing results are recorded as part of quality control. The electronic recording of the welding parameters is helpful, but cannot replace the tear tests.[1]

The heat development during welding produces shrinkage of the weld seam in the long direction. This shrinkage, also

Fig. 114: Welding under pretension

called **thermal crumple**, is compensated during the welding process either by pretensioning the fabric (Fig. 114) or it has to be compensated in a calculation and pulled out of the fabric during erection. The thermal shrinkage must always be taken into account when cutting out.

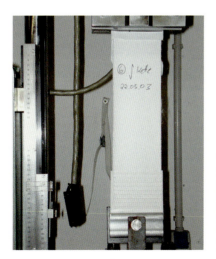  

Control test   Test of the completed seam   Torn seam

Fig. 113: Seam tests

1 Rudorf-Witrin, W. (2004)

Fig. 115: PTFE heat welding press with a heating beam length of 2 m and a maximum pressure of 7 bar

Hot element welding

Strips of PTFE-coated fabric can also be welded by heat contact welding, a special form of heating element welding. The coating masses are heated through contact with a heating beam to a temperature of up to 340 °C. PTFE is composed of linear chains and becomes thermoplastic when heated. After the crystalline areas have melted, it is not sufficiently liquid to be processed further. To get round this, a layer of thermoplastic foil is trapped between the fabric pieces to be joined as a welding aid. Then the pieces to be welded are pressed under a pressure of 50 N/cm² for 30–40 sec. On cooling, a chemical bond is formed between the coatings. This type of joint can reach 80–90 % of the fabric strength.

ETFE foils can be welded together thermally using a welding beam or by heat impulse welding. In neither case is a welding aid used. The welding by contact with the beam can proceed continuously or cyclically. Welding is normally done at temperatures above 230 °C.

The crystallite of partially crystalline plastics melts at welding temperature, so the polymer is then available as melted mass. On cooling out of the melted mass, the polymer crystallises again, the density in the crystalline areas increases and the volume reduces. To achieve the required quality of weld seam, rapid cooling of the seam is required after heating. The parameters for the cooling conditions and the tool temperature are according to the experience of the fabricator and have a major influence on the material properties.

The seam widths for cushion applications, depending on the loading, are normally 5–20 mm. According to information from the manufacturer, the strength of the seam is about > 90 %.[1]

In thermal impulse welding, the welding heat is produced by electrical impulse in a thin metallic welding band through resistance warming. The welding temperature can be regulated precisely. High seam strengths can be achieved by subsequent cooling under pressure. The advantage of this process is the rapid cooling of the heating bar.

Fig. 116: Welding of ETFE foils

[1] Fitz, H. (2004)

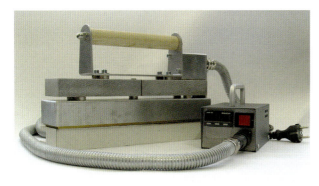 

Fig. 117: *left:* Manual welding device; *right:* Mobile magnetic welding device

Welding work on site

For waterproofing and for touching-up and repair work to damaged fabrics, welding must often be carried out on the construction site. Various hand welding tools are used for this purpose.

The site welding of covering membranes made of Glass/PTFE fabrics is now mostly done with hand devices, which apply the pressure onto the fabric with a hand iron. The fabric pieces to be joined, with the welding aid, are welded together at a temperature of about 360–420 °C under pressure. This process lasts about 1–2 minutes per weld. Devices in use today have a temperature control for a range up to 450 °C (left in Fig. 117).

To weld Glass/PTFE fabrics, magnetic welding devices can also be used. These devices work at a welding temperature of up to 390 °C, are equipped with a heating beam and have a similar pressure to stationary welding presses. They do, however, weigh about 50 kg and are therefore only usable for mobile site welding if hung from a crane (right in Fig. 117).

Waterproofing and repair work on PVC-coated polyester fabrics are usually performed with a hot-air pistol (left in Fig. 118), a welding process that was also formerly used for fabrication. It is possible to work at a temperature range of 50–600 °C with the hand devices, which weigh approx. 750 g. The welding of Glass/PTFE fabrics with a hot air pistol is indeed possible, but is seldom done on account of the highly poisonous vapour given off by the welding process.

Joints in partial surfaces between fabric membranes are normally made by the fabricator in advance. If the area to be roofed over cannot be delivered in one piece, then assembly joints have to be provided in order to join the adjacent parts.

For the erection of the vehicle park roof at the Munich Waste Management Office, a high point structure, a conscious decision was made to use stiff clamping connections for the assembly joints between the partial areas. The prefabricated panels approx. 10 x 12 m in area of the 8.400 m² Glass/PTFE membrane were welded at the works in 70 m long strips and delivered to the construction site, where they were laid out and welded together. In order to be able to better compensate errors, the site weld seams were made with double width (right in Fig. 118).[1]

Fig. 118:
left: Repair work with a hot air pistol;
right: Welding of partial surfaces of Glass/PTFE membrane at the Munich Waste Management Office (AWM), Germany

1 Göppert, K. (2003)

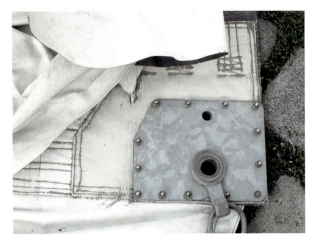

Fig. 119: *left:* PES/PVC fabric corners of a circus tent reinforced with sewn seams; *right:* Sewing the edge area of a Glass/PTFE fabric

2.5.1.2 Sewn seams

Sewn seams, which are the traditional means of joining fabric in tent building, make possible a direct connection of fabric thread to fabric thread to transfer force. In the construction of lightweight structures, however, sewn seams to join two sheets are rather the exception today, particularly as the perforation of the membrane by the sewing needle damages the waterproofing and has to be waterproofed later.

In highly exposed edge areas and corner cut outs, sewn seams are still used today for connections, with the welded seams in these areas being additionally sewn (left in Fig. 119). When edge belts of Glass/PTFE fabric are used, these are also sewn to the fabric (right in Fig.119).

When sewing a seam, special care has to be taken of "tidiness". Excessive sewing speeds heat up the sewing needle strongly and burn holes in the fabric. Known forms of sewn seam are flat seam, turned-in seam and hem seam. The stitching types are lock stitch, zigzag and warp stitch. The best seam, but also the most expensive to produce, is the double turned-in seam, which has the two hems hooked over one another and is then sewn with many parallel lines of stitching. In order to provide weather protection and stop light passing through, the sewn seam can be welded over with a foil and sealed. This is normally done on the construction site by welding a prefabricated cover flap over the sewn seam.

2.5.1.3 Glued seams

Glued joints are only used in membrane construction today to join the seldom-used silicone-coated glass fibre fabrics. This type of fabric cannot be welded on account of its structure. Silicone is an elastomer, whose molecules are not worked on a large scale, and elastomers cannot plasticize. If they are to be joined, then they have to be vulcanised similarly to rubber. The rubber elasticity is also preserved down to low temperatures.[1]

In order to be able to glue the part to be joined, the surface of the silicone coating must be treated. The solvent in the adhesive swells up the surface, the adhesion of the molecules is broken and a better adhesion is reached. Finally the adjacent pieces can be glued together with the networked adhesive. The strengths, which can be achieved with the correct parameters for the networking process, are comparable with the material strength.[2]

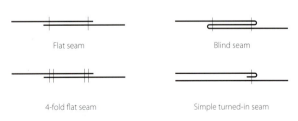
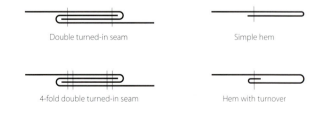

Fig. 120: Types of sewn seams

---

[1] *DVS guideline 2225 (1991)*

[2] *Blum, R. (2002-2)*

2.5.2 Reusable surface joints

Because of the limitation of available strip widths, membrane surfaces, which cannot be transported and erected in one piece, are detailed with reusable erection joints. Such joints can be bolted or laced, depending on the demands of force transfer, details and practicality of erection. High forces have to be transferred out of the relatively very thin membrane sheet through the bolted or laced joint into the adjacent sheet. On account of its construction with metal plates, the reusable joint represents an irregularity in the stiffness distribution of the membrane surface, where obstructions to the overall deformation of the membrane occur.

2.5.2.1 Clamping plate, keder rail joints

The most important reusable joints for the transfer of high forces are clamping plate joints and keder rails. The clamping plate joint is a combination of frictional and form-fit connections; the fabric strips have a keder rail and are clamped on the construction site with bolts between two aluminium or stainless steel plates. This joint is also suitable as a fixed edge anchor of the membrane to the primary structure. It can be applied as a flat, valley or ridge joint.

The force is transferred through a keder installed at the edge of the metal plates each side, the purpose of which is prevent the sliding of the membrane out of the clamping plate. The strength results for the clamping plate joint are similar to welded seams. If the load transfer of the clamping plate joint is entirely into the plane of the membrane, then the length of the plate must be suitable for the surface curvature, in order to avoid the formation of folds between the plates.

One problem with joining panels with clamping plates derives from the different stretching properties of the membrane inside and outside the bolted joint. The membrane therefore has to be partially prestretched during erection before bolting the clamping plate in order to distribute the calculated compensation along the whole length of the joint and the spacing between the clamped profiles.

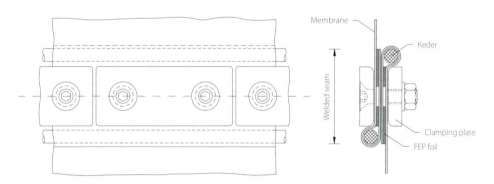

Fig. 121: Clamping plate butt joint

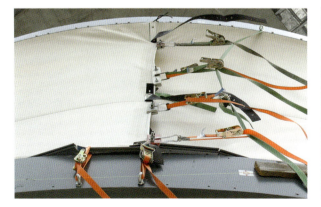

Fig. 122: *left:* Assembly of a clamping plate but joint; *right:* Stretching the membrane in a keder rail

Fig. 123: Prestretched erection joint in an arched surface

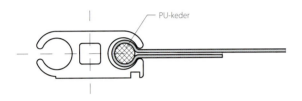 

Fig. 124: *left:* Keder rail for fabric membrane; *right:* Clamping plate for ETFE cushion joint

As an alternative to the clamping plate joint, preformed keder rails made of aluminium of plastic can be used for joints between partial sheets. One and two-part rails are available, which can accommodate 1 – 4 keders. The poor sliding capability of the keder hem in the keder profile can be a problem during erection (right in Fig. 122). In order to prevent the fabric coating sticking in the profile, special non-stick keder strips can be fabricated onto the keder.

2.5.2.2 Looped and laced joints

For less heavily loaded joints, as are used for temporary structures or large pneumatic cushions, looped or laced joints are often used.

Polyester cords are mostly used as laces. The two-part eyelets are mostly made of galvanised or stainless steel and consist of a trumpet and a washer. They are welded into the membrane in one step using a welding/stamping press, with the opening being stamped out at the same time. The diameter and length of the laces and the shape and diameter of the eyelets are determined depending on the loading. The force transfer is from the keder through the eyelet or clamping plate holes to the lace, and it is important to arrange the eyelets close to the keder.[1]

The tensioning during erection is performed in stages of tightening and loosening the laces. The variation of the opening width by tightening or loosening the laces enables any imprecision in the cutting out of the pieces to be compensated. The advantage of tensioning in stages does, however, result in increased working time. Because of the good flexibility for lateral adjustment, stretching of membrane and laces can both be easily compensated.

The unprotected joints can be covered with a membrane strip welded or sewn on one side and secured with a Velcro or buckle fastening.

Fig. 125: *left:* Laced connection of the large pillow of the arena in Nimes, France; *right:* Loop joint of a circus tent

1 *Bubner, E. (1997)*

2.6 Methods of transferring force at the edge

Membrane surfaces are curved on one or two sides within a closed edge. In order to stabilise such surfaces in their intended form, forces are introduced through linear bearing elements at the edges.

2.6.1 Geometry of the edging and effect on bearing behaviour

The bearing and deformation behaviour of a membrane surface is essentially determined by the geometry of the edging. Edge details for membrane surfaces can be stiff or flexible in bending. The tension forces are introduced through flexible edges entirely as tension, whereas stiff edge elements are mainly loaded in compression and can also be subject to bending moments.

The detail of the edging, and also the forces acting on the edge element, influence the geometry of the surface and thus the stiffness of a membrane. Loading causes deformation in the membrane, which a flexible edge can resist by deformation of the edge element. The relaxing of the edge element causes a relocation of the supports of the membrane surface. The resulting deformation of the edge leads to a reduction in the maximum stress.[1] This can be advantageous for the structure, because it may be possible under favourable conditions to save tensioning force and concrete for the anchorage.

With a stiff edge detail, the membrane has no possibility to relax peaks of stress through deformation of the edge. Apart from the higher weight of the edge element, a design with stiff edging can in some cases require a stronger type of material for the membrane surface. The higher strength of the material then leads to increased geometrical stiffness of the membrane surface and, on account of the higher forces to be introduced, to a slightly altered surface curvature in comparison with the option with flexible edge.

A freely curved, flexible edge normally lies in one plane. Any transverse deformations (deflection under external loads) produce an alteration of the curvature in the surface. The force in the flexible edge element, the radius of the edging and the membrane stress are factors, which can influence the surface geometry of the membrane in the following way:

The membrane stress (SM) increases with increasing force in the edge element (SS). The relationship of edge rope force to membrane stress determines the edge geometry. An increase of edging radius (R) produces a reduction of the surface curvature, which means that increasing curvature radius leads to increased forces in the edge elements (Fig. 127). In order not to exceed the maximum stress in the edge element, the dimensions of the edge element have to be designed for forces.

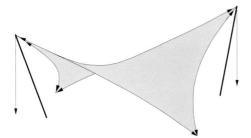
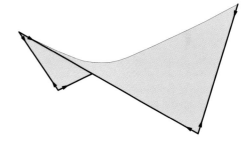

Fig. 126: Flexible and stiff edging of a four-point awning

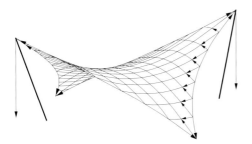
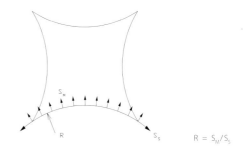

Fig. 127: Membrane form as a function of the edge geometry

---

1 Essrich, R. (2004)

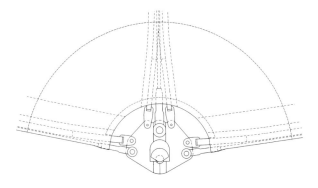
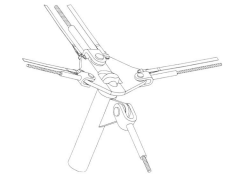

Fig. 128: Form-fit, friction and bonded connection at a membrane plate

2.6.2 The detailing of edges and their anchorage at corners

Because flexible and stiff load-bearing elements work together to transfer loading, the deformations of the individual load-bearing elements have to limit each other and be "compatible" with each other. This results in an important parameter in the detailing of edge and corner details – the consideration of the stiffness relationship between membrane and edge reinforcement. The difficulties in detailing this arise from the need to introduce high tension forces from thin, high-strength and flexible membranes into the low-strain, stiff metallic elements, and to absorb differences of deflection. Near the edge, the deformation of the load-bearing elements has to be compensated in consideration of the different deformation behaviour of the materials. To reconcile the different stiffnesses in strain and bending of the elements to be brought together, a combination of force-transfer, bonded and form-fit connections can be used.

If the edge element and the surface element consist of the same material, then the weight ratio of surface to edge element with changing radius can be used to derive a relationship to the corresponding stretching of the two materials.[1] Deformation requirements on the detailed design and choice of materials for the edges and their anchorages result from the conditions of surface and edge geometry for the capability of resisting tangential forces resulting from translational and rotational deflections under loading. This applies especially during erection, when the change from loose to tensioned condition causes large changes of form.

The rotational angle of the fixing of the membrane at the edges and corners should be taken into account in the specification of the degrees of freedom at the anchorage points. If one direction of the fabric becomes slack, the angle of rotation at the bearing as a result of wind and snow loading can increase considerably. During erection, the direction of introduction of the force during the tensioning process particularly plays an important role in tensioning the membrane without folding or breakage of the coating, particularly for PTFE-coated glass fabrics, which are at risk of kinking.

The direction and extent of the incoming forces also need to be taken into account in the design of the constructional details. Radial forces with different directions and sizes act on the edge depending on the size of the pretension introduced in warp and weft direction, or in bearing and tensioning directions. If a fabric strip meets the edge at an angle, then loading will cause a different effect of the edge element, which must be absorbed to avoid the fabric sliding (see Fig. 129).

A further aspect in the design of edge and corner connections is their dimension. If a structure is meant to be built as light and slender as possible, then the details should al-

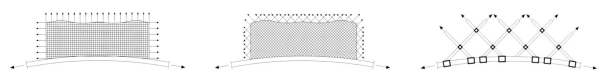

Fig. 129: Forces acting on the edge element

---

1 Otto, F.; Happold, E.; Bubner, E. (1982)

so be designed to function with a minimum of weight. The connections, however, can only be reduced to a minimum size, taking into consideration the space requirements of the various load-bearing elements coming together and their anchorages.[1] This limit to the adaptability of the details to the size of the construction also has to be observed when designing details, which are practical to erect.

2.6.3 Edge details

Forces out of the membrane surface have to be transferred through the edging into the primary construction. Various types of edge elements have been designed to do this. The main design factor is how the forces are to be channelled and introduced into the primary construction.

If an edge element runs along the edge of the membrane that collects the forces from the membrane in a curved line and leads to the anchorage, this is called a *flexible edge detail* (a – d in Fig. 130). A load-bearing element tangential to the edge of the flexible edge can only be loaded in tension, which is uniformly distributed over its cross-sectional area.

If the tangential forces are transferred to linear, fixed or multi-part formed components and introduced from there into the adjacent structure at a point or along a line, then this is called a *stiff edge detail* (e – h in Fig. 130).

Further design parameters for the development of edge details are the stiffness relationships and the deformation behaviour of the different materials, the dimensioning with regard to an economic balance between span and fixings, the practicality of erection and the creep behaviour under biological, chemical and physical influences.

In the following section, the most commonly used flexible and stiff edge details are described. The practical question of erection is also discussed.

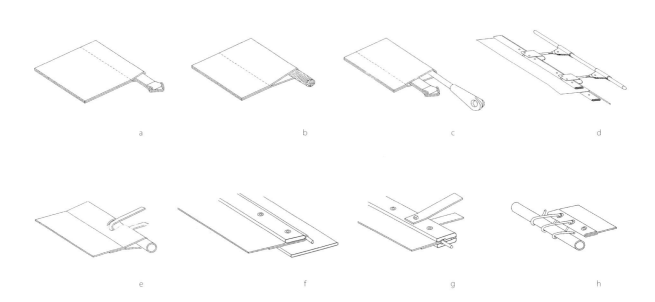

Flexible edgings
a   webbing edge
b   rope edge
c   rope edge with webbing
d   clamping plate hung from rope

Stiff edgings
e   tube edge
f   clamping plate edge
g   clamping plate edge with brackets to stiff edge beam
h   tubular edge with lacing

Fig. 130: Membrane edge details

1   Sobek, W. (1994)

2.6.3.1 Details with flexible edging

One commonly used flexible edge detail is the rope edge, where a rope running in a sleeve supports the edge forces. The edge is freely spanned and curved in an arch.

The sleeve is formed by folding over and welding the edge of the membrane. This forms a hollow hem. During erection, a steel rope is pushed through as edge element (Fig. 131). In order to avoid damage to the hem, there is usually abrasion protection fitted inside. Edge rope sleeves can be mounted as separate sleeves like a sandwich. If they are only welded to the membrane on one side, this leads to considerably lower strength results.[1]

The width of the hollow hem depends on the tension stress perpendicular to the edge. The hem width and the rope diameter determine the extent of the splay of the hollow hem (α in Fig. 131). When loading is applied to the hollow hem, the coating of the fabric is loaded axially. The splay produces a combination of peeling and shear loading at the weld. With increasing splay angle, the proportion of peeling action increases with tension load; this can lead to the peeling of the coating surface from the fabric.[2]

Uniaxial and biaxial tests can be performed to test how the peeling strength varies under biaxial loading and at various angles of the fabric direction to the edge. Experience in practice is that the limitation of the splay angle to 15° avoids the weld seam peeling off. For rope edgings of PTFE-coated glass fibre fabric, the maximum angle should be 6°.[3]

If the friction forces between rope and sleeve are not sufficient to resist movement between rope and membrane, then additional webbing to absorb these tangential forces can be mounted in the area of load introduction (Fig. 132). The webbing, mostly polyester, can be compensated or pretensioned before fixing the membrane, so that the webbing can accept the strains produced by pretensioning.

For the transfer of these forces, the bond strength between webbing and membrane is decisive. If the forces to be introduced are relatively low, sewn-in or welded-on webbing can be used alone. Polyester webbing has, however, only limited strength, so webbing edge details are normally only used for membrane surfaces with relatively short spans.

The edge can also be constructed to be stiff in bending, with a tube being used as tensioning element. Edge details with tubes running in sleeves are quick to erect and relatively cheap. Fixed to the ground with ground anchors, this was for a long time the standard practice for anchoring smaller air-supported halls. This solution is not suitable for high loadings.[4]

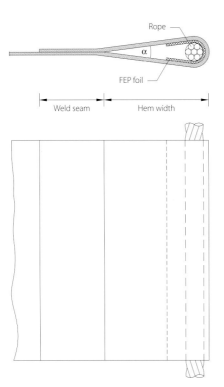

Fig. 131: Hem sleeve with rope as edge element

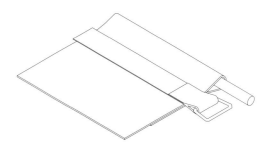

Fig. 132: Rope edge with sewn-on webbing

1 Stimpfe, B. (2000)
2 Minte, J. (1981)
3 Stimpfe, B. (2000)
4 Sobek, W. (1994)

If it is not desirable to run the rope inside the sleeve, then the rope can also be laid externally. The forces from the membrane then have to be transferred from the keder to the clamping plate and from there onto the external rope through sheet metal loops.

Such an edge detail with a combination of rope and clamping plate is often used for membrane surfaces with a large span (Fig.134). The solution is often constructed as flexible edging with PTFE-coated glass fibre fabrics. A rope edge detail as used with PES/PVC fabrics is less common with glass fibre fabrics, because the PTFE coating peels off easily. The occurrence of high spreading forces at the sleeve must be prevented.[1]

The detail shown in Fig.133 has the fabric held in a linearly clamped keder edge and hung from the rope with sheet metal loops at regular intervals (a in Fig. 133).

A continuous, linear form-fit between keder and clamping plate is only achieved when the spacing between fixing bolts is not too wide and uniformly spaced (b and c in Fig. 133). The same applies for the separation of the clamping plates (d in Fig. 133).

The connection is freely rotating at the edge, but the introduction of forces parallel to the edge is to be avoided on account of the slanting produced. A better distribution of forces in the longitudinal direction can be achieved with an arrangement on the underside (u in Fig. 133).

The cost of erection of the loop edge is considerably higher than for the rope edge, where the rope can be relatively simply pulled through the sleeve with a guide rope. With the loop edge, the individual plate top and bottom parts must be preassembled onto the membrane, which has partially to be prestretched longitudinally, and then each loop fixed onto the rope.

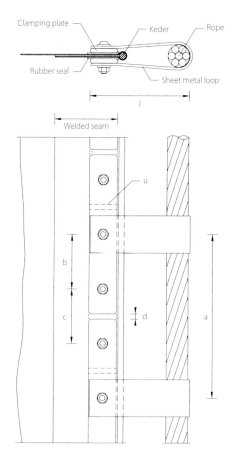

Fig. 133: Edge detail with sheet metal loops

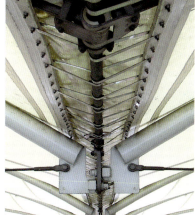

Fig. 134: *left:* Erection of valley for the roof over the Sony Center in Berlin, Germany; *right:* Valley detail at the Gottlieb Daimler Stadion in Stuttgart, Germany

1 Stimpfe, B. (2000)

Fig. 135: Tensionable edge detail of the Grande Bigo hall in Genoa, Italy

A tensionable edged detail was used for the Hall Il Grande Bigo in Genoa, Italy, erected in 1992. This has the edge of a PTFE-coated glass fibre fabric held in clamping plates, which are tied to the edge rope by shroud tensioners, with a specially manufactured fork holding two clamping plates (Fig. 135).

If weak forces are to be led along the flexible membrane edge, then this can be detailed with a webbing edge, which can be single or double sewn webbing. These can be in a sleeve or externally welded or sewn to the membrane (a, b in Fig. 136). In either case, they are connected firmly to the membrane. Webbing in the open must be protected from UV radiation and biological effects, which is done by wrapping membrane material around.

Because of the difficulties involved in coordinating the strain stiffness of the membrane and edge reinforcement to each other, Hans Gropper and Werner Sobek have proposed as a new development a membrane edge element. This starts from the premise that edge elements, which are weak in strain, result in more favourable membrane force conditions between membrane plate axial and transverse forces in sharp corners than, for example, steel ropes. This design for a textile edge element with asymmetrical cross-section was displayed at the trade fair Techtextil in 1985 in Frankfurt am Main.

It consists of polyester (PETP) fibres woven together with a total breaking load of about 1,000 kN and can be connected with the membrane fabric using conventional fabrication methods. The connecting fabric transfers radial and tangential forces into the thicker section part of the edge element, which has a load-bearing function. After a prototype had been made, tests were commissioned on possible anchorages for the textile edge element. These were sliding mandrel and a cast plastic anchorages.[1]

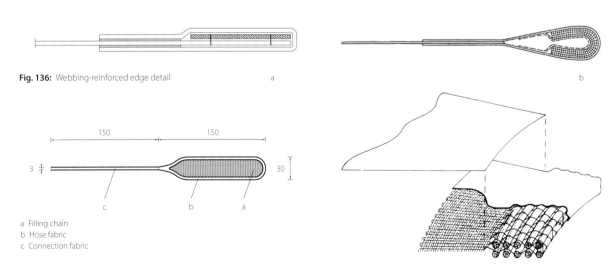

Fig. 136: Webbing-reinforced edge detail

a Filling chain
b Hose fabric
c Connection fabric

Fig. 137: System diagram for a textile edge element from Hans Gropper and Werner Sobek

1 Gropper, H.; Sobek, W. (1985)
   Remark: No membrane structures with this type of edge detail are known to the author

2.6.3.2 Details with edges stiff in bending

The most important reusable connection for the transfer of high forces is the clamping plate joint, also called a sectionalising joint. With this combination of friction and form-fit connection, the fabric strips to be connected have a keder and are pressed between two metal plates on the construction site. The connection is suitable for the fixed edge anchor of the membrane to the primary construction.

The force transfer is through the edge of the keder situated at the edge of the metal plate, whose purpose is to prevent the membrane from slipping out of the clamping plate. If clamping plates are selected as edge detail, then the membrane has to be perforated. Care has to be taken here, because the spacing of the holes for the bolts is altered by prestretching. A continuous, linear load transfer between membrane and primary structure is only ensured when the bolt size and the hole diameter in clamping plate and fabric match each other. If the clamping plate cross-sections are not sufficient, the edge keder could be pulled out from under a bent clamping plate.

The form-fit load transfer from the clamping plate through the keder to the fabric thread will only function correctly when the keder is laid directly at the edge of the clamping plate during erection. If this is not ensured, then tearing of the membrane could start from the stamped hole because of slack in the side of the hole. The holes punched in the membrane for the clamping bolts should be so large that the edges of the hole in the membrane do not touch the bolts (a in Fig. 138). This is especially important for glass fabrics.

The length of the clamping plates is according to the rule of thumb: the less the curvature of the membrane, the longer the plates can be. The stretching of the membrane in the long direction should not be obstructed by the clamped connection.

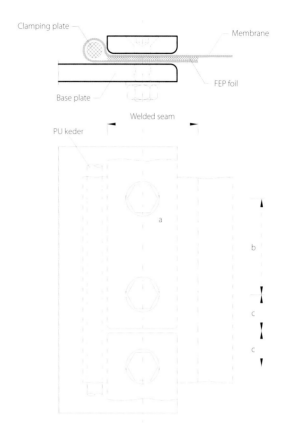

Fig. 138: Clamping plate edge

Plate lengths up to 150 mm are usual, with shorter plates being safer for more pronounced curvature. For clamping plate edges, which are stiff in bending, lengths up to 1,000 mm are used.[1]

The nuts of the clamping plates are tightened with a defined torque setting. When the erection takes place at low temperatures, the level of tensioning force should be checked to avoid brittle fractures of the metal plates. The torque of

Fig. 139: Clamping plate edge and corner detail for the roofing of the Main Station in Dresden, Germany

1 Bubner, E. (1997)

Fig. 140: Keder rail profile

Fig. 141: *left:* Corner connection and edge detail to the roof over the entrance to the Federal Chancellor's Office in Berlin, Germany; *right:* Detail photo of clamping profile

the clamping bolts has no influence on the strength of the connection if the assembly is correct. The spacing of the bolts does, however, have an influence on the strength of the joint (b in Fig. 138). The spacing should be hardly more than 200 mm.[1] The clamping plate dimensions should be designed so that when the membrane is strongly curved, the plates cannot come into contact (c in Fig. 138).

Stiff membrane edges can also be constructed with the single and multi-part profile rails mentioned in sections 2.2.1.3 and 2.5.2.1. When the edge detail includes a keder rail, the ability of the fabric to slide through the rail is important.

Stiff edge versions normally represent a friction-fit system with the primary structure. The edges are mostly straight. If the edge is curved, then the practicalities of manufacture mean that this will normally be curved in one axis.

One special form of the stiff edge detail was used for the 240 m² roofing over the entrance to the Federal Chancellor's Office in Berlin.

The radial and tangential forces from the PTFE-glass fabric are carried here through three-dimensional clamping profiles of aluminium gripping ropes connected to the primary structure. The curved profiles are connected to each other with internally bolted linking plates to ensure the required longitudinal stiffness. The edge keder of the fabric is formed as a round strand rope.

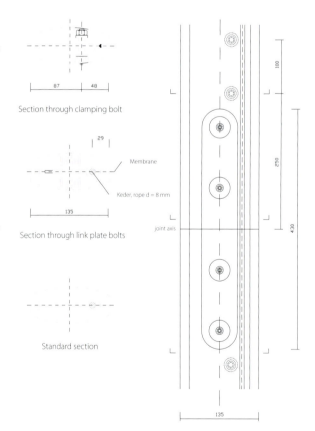

Fig. 142: Detail of an aluminium clamping profile

1 Minte, J. (1981)

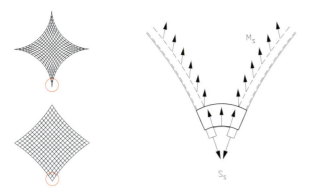

Fig. 143: Schematic illustration of the possible course of forces

2.7 Corner details

Membrane corners form the end point between two membrane edges, where the ends of the load-bearing elements are anchored to each other through a metal fitting without relative movement being possible. At the corner, the forces from the membrane have to be diverted through fittings and transferred to the primary structure. The changeover from unstressed to stressed condition through the application of pretensioning makes the corner area particularly at risk from overstress and the formation of folds.

The edge detail and its anchorage at the membrane plate influence the load-bearing behaviour of membranes considerably. This is clear from the direction of the membrane forces to be resisted and also through the stiffness relationship of edge reinforcement and membrane.

For membrane edge details, the extent and direction of the forces to be resisted are also dependant on the geometry of the corner, the angle in the membrane plate area of the membrane surface. The edge element in the membrane plate area runs almost parallel to the main curvature of the membrane surface. The membrane stresses, which near the edge run axially to the edge rope tangent, are introduced in the membrane plate almost parallel to it. The more acute the angle of the corner is, the more important is the capability of resisting

Fig. 144: Corners with edge rope with obtuse and acute angles

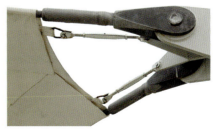

Fig. 145: *left:* Webbing corners with continuous and butted webbing; *right:* Rope corner with tangential webbing

Fig. 146: *left:* Rope corner with lattice membrane; *middle:* Rope corner with webbing reinforcement; *right:* Rope corner with tangential webbing

membrane forces acting parallel to the edge curve, in order to prevent the membrane sliding out of the membrane plate.

This can be achieved with a flexible edge detail through the provision of corner fittings. Depending on the sliding force and the edge geometry, this can fix the membrane without movement being possible (Fig. 144).[1]

Corner fittings and edge elements do, however, represent stiff components at the corner. This leads to an undesirable stiffening of the corner at the membrane plate. The more acute the angle in the membrane plate, the higher are the membrane stresses to be expected and the larger is the influence of the edge element in making the corner stiffer.

If the stiffnesses are coordinated through the compensation of possible deformations between corner fitting, edge element and membrane, then this leads at the membrane plate to an unfavourable relationship between transverse tension and longitudinal forces in the membrane plate, which at least poses a risk of folds forming.[2]

If only low forces are to be resisted through the edge, then textile edge elements can be continued to the corner and anchored by metal plates or rings (left, centre in Fig. 145). Such details do indeed improve the stiffness relationships considerably, but the risk of folds forming is still present.

Because of the need to transfer forces, flexible edges with ropes will still be detailed in most cases. This does, however, represent a problem for the tensioning process during erection on account of the stiffness relationships between stiff edge element and corner fitting. At this location, the forces in the membrane have to be influenced by the introduction of the pretension so that an angular rotation can take place in the fabric.

If the corner area is exceptionally acutely angled, then the limiting angle, at which the resistance to this rotation increases, becomes smaller and the geometrical stiffness in the fabric increases. This has the result that higher forces must be introduced in the membrane plate area during pretensioning than into obtusely angled corners.

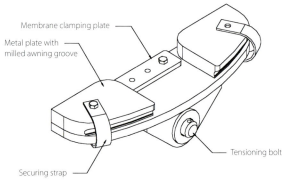 

Fig. 147: Edge fitting with continuous rope

1 Bubner, E. (1997)

2 Sobek, W.; Gropper, H. (1985)

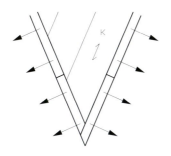 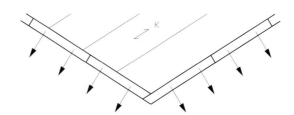

Fig. 148: Stiff corners with acute and obtuse angles

If the edging is flexible, then the pretension is mostly applied by pulling the corner fitting. Fabricated membrane strips or webbing can raise the material capacity in this critical area if unusual forces have to be introduced (Fig. 146). This measure does however increase the stiffness at the corner. This has to be taken into account in the determination of the level of the compensation values and in the planning of the pretensioning procedure.

Considerably fewer problems have to be feared during pretensioning at obtuse, flexible edges. In this case, the rope can be continued round the corner fitting (Fig. 147). This also makes any later tensioning adjustment of the compensated edge element easier.

For stiff edging, the strain stiffness of the edge element results in a relationship with the direction of the planned strip layout during the pretensioning process. It is not possible to tension the entire edge uniformly here because of the movement of the corner fitting. The introduction of the pretension is done by pulling the corner element in stages. The cutting-out direction arriving in the corner area does, however, not correspond to the axis of the pretensioning force to be introduced. The introduction of higher forces in the main anisotropy direction also makes tensioning considerably harder if the corner detail is stiff in bending, particularly at acutely angled corners (Fig. 148).

If the strips are cut out parallel, the necessary strain for the intended pretension varies along the edge curve. But the strain stiffness of the element is constant over the entire length. The resulting stress differences in the membrane have to be resisted in the membrane plate area.[1]

These factors make it clear that the cutting out and the seam length play an important role for the membrane plate area and make careful consideration of the design of these details necessary. The corner area of stiff edges is mostly detailed as an alteration of direction in the stiff edge element. These normally consist of standard clamping plate elements (Fig. 149).

Fig. 149: Stiff corner details

1 Gropper, H.; Sobek, W. (1985)

3 Construction of tensile surface structures

3.1 Introduction

In addition to the design of the structure, design of cutting-out and details, and the production of the materials, the construction is the most important stage in the chain of processes leading to the implementation of a membrane structure.

The construction process serves to erect the membrane structure at the intended location. Single components are assembled to form a load-bearing structure according to the drawings and specifications. Construction includes all operations intended to bring together non-demountable parts to larger groups or elements (assembly units) and the installation or erection of individual parts or assembly units.

The building components must be assembled, checked, positioned and adjusted through appropriate handling. The construction is complete when the specified strength of the joints and elements has been achieved at the intended location in the structure. This is the case with membrane structures when the structure and its load-bearing elements have assumed the intended geometry through the application of the calculated pretension.

The production and preassembly of the individual components to units suitable for transport is done at the works. Further preassembly and all operations required to erect the structure take place on the construction site, making use of the necessary temporary devices and equipment and to a coordinated schedule. The construction operations can be categorised into subsidiary works, operations to implement the assembly and processes of control and documentation.

The major advantage of membrane erection is above all the short construction time. High precision of fit and excellent quality can be achieved, supported by production in highly specialised companies. The typical characteristics of the erection of membrane structures are the specialised manual and mechanical handling of the flexible structural elements on the construction site. The purposeful material combination of the edge and sheeting elements produces a complex mechanical procedure, which should be taken into account in the planning and implementation of the erection.

The inadequate stability of the load-bearing elements before they are tensioned, for example, demands the creation

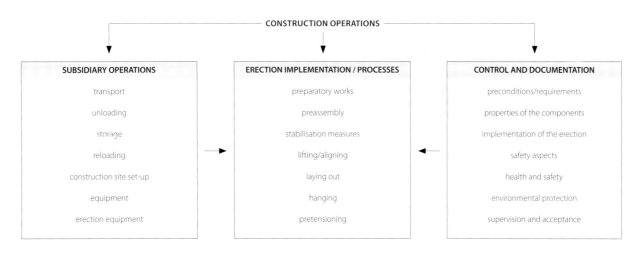

Fig. 150: Assembly operations – chain of processes

of temporary loading cases for the structure during erection, which no longer apply in the pretensioned condition. The implementation of the necessary stabilisation measures demands exhaustive consideration in the design of the individual components. Structural calculations for the load-bearing elements and for the entire structure also have to be carried out for assembly stages, and junction and fixing points to provide temporary anchoring for the unstable elements have to be provided.

Particular attention also has to be paid to the effects of time and temperature on the materials during construction, because the effect of these on various types of fabric while applying the loading can be very different.

Because erection takes place at the end of the project schedule and there is normally little time available for alterations, any errors or deficiencies in the design or production process will become evident at this point. This needs to be taken into account early in the project schedule through the corresponding determination of tolerance specifications.

The contents of the present chapter

The present chapter is essentially structured in parallel to the sequence of actions during a project to erect a membrane structure.

Starting with an overview of the role and tasks of construction management, the various areas of construction management and their economic and technical aims are summarised. In additional to the constraints of time and technology for construction management, the modelling of construction processes and the optimisation of design detailing for ease of erection are also discussed.

A description of the erection equipment in use is intended to give a summary of devices and tools for transport, lifting and tensioning. The emphasis here is on equipment for the tensioning of flexible structural elements.

In the next section, the principles underlying the construction process and the factors influencing them are laid out. Particular attention is paid to the working principle of the structural system regarding its structural form and stability while being erected. Subsequent descriptions of the construction of the primary structure include details of the construction of masts, edge beams and rope structures.

Then there is a description with examples of the procedure for the erection of characteristic structural forms of mechanically tensioned membrane constructions. Single stages in the construction process are illustrated and the erection schedule correspondingly commented. Finally, there is a picture gallery of completed projects.

A further section is dedicated to the implementation of the construction and the procedures on the construction site. From the preparation work and the preassembly through lifting, hanging and tensioning, the essential processes for the erection of structural elements are illustrated with examples. A particular emphasis is on the problem of introducing the forces into the membrane surface.

In the last section, there is a final summary of methods and procedures for the measurement of forces in ropes and membranes.

3.2 Construction management

The planning forecast of the construction process and the economic and technical preparation to coordinate the work on the construction site are described in the building industry as construction management. The purpose of construction management is to optimise the construction period and the deployment of personnel and equipment for the erection of a structure.[1]

Despite a high degree of prefabrication, the demand for ever shorter construction times is also being experienced in the field of lightweight structures. Complex load-bearing structures and the application of ever more highly developed materials mean that the importance of construction management is becoming more evident. To realistically estimate the practicality and the relation of production cost to erection cost for a membrane project is a matter of experience. Experts in fabrication and construction will also need to be consulted during planning.

Construction management consists of complex areas of economic responsibility. In addition to the commercial tasks of the quantity surveyor, the construction manager also faces challenging technical demands for the production of erection schemes. Particularly in the field of membrane construction, where a multitude of special structures are erected with ever more varied solutions, the knowledge of the production and assembly process plays a decisive role in the optimising of erection cost and thus the achievement of shorter construction programmes. This makes clear that efficient construction management is very relevant to the cost of erecting a membrane structure.

3.2.1 Aims and tasks of construction management

The preparation for the erection of lightweight structures involves many areas of work and has to implement measures in the fields of design, production, transport, assembly and supervision to ensure the trouble-free course of construction progress.

These measures include the areas

- *Design* – the production of a construction programme and detailing for ease of construction,
- *Production* – to suit assembly methods,
- *Delivery* – Delivery of materials at the right time according to construction schedule,
- *Assembly* – unhindered technical construction progress on the site and
- *Control* – the associated measuring, checking and documentation.

Construction management places heavy professional requirements on the responsible engineer. Foresight and organisational ability as well as thorough technical training are essential in order to be able to cope with the complex task. The classical areas of responsibility of the construction manager include the commercial management and the technical supervision of the project.

The areas of responsibility of construction management can be summarised as

- Scheduling of progress,
- Capacity planning and resource management
- Technical erection planning.

The responsibility for construction management mostly follows from the size of the project and the complexity of the tasks. The larger and more complex the project, the more likely it is that an experienced engineer will be responsible for construction management. With smaller, more simply constructed membrane structures, the construction management and erection are normally undertaken by the construction firm or the fabricator.

If the construction management and execution are the responsibility of the fabricator, then there is in effect no risk of information deficiency between production and erection. Production and construction management combined in one house also have the advantage that production imprecisions can more easily be rectified during erection.

Textile architecture only represents a very small portion of the construction market. This has the result that there are very few specialised erection companies. They are, however, mostly very well equipped and possess wide experience in project development and implementation. They are often very flexibly structured as regards their personnel capacity, which is particularly advantageous for overseas contracts.

The planning of the construction of larger-scale projects is mostly the responsibility of engineers. The structural engineer usually collaborates with specialised consultants or assigns the task completely. On the construction site, the construction design team accompanies all stages of erection, beginning with the positioning of the anchorage elements and continuing with the construction of the structure, then the erection of the membrane and the mechanical or pneumatic application of the specified pretension, and finally acceptance and handover.

---

1   Albrecht, R. (1973)

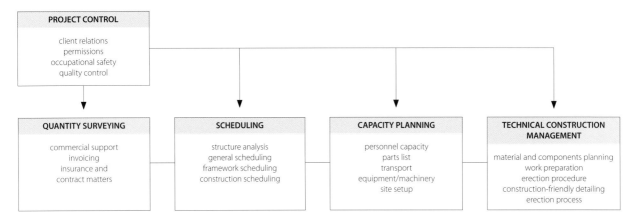

Fig. 151: Tasks in construction management

The personnel entrusted with the construction must be sufficiently qualified for the preparation and assembly work on site as well as rope-supported work, often at great heights. The personnel, who mostly come from steel erection, must be adequately trained in handling and working with flexible load-bearing elements. The valid legal regulations concerning construction also have to be observed.

3.2.2 Scheduling

The planning of the time required for design, production and erection is one of the main tasks of construction management. It serves to organise the individual work sections and determine when they should be performed, and delivers the background information regarding capacity and cost planning for the deployment of personnel and machinery.[1]

With the coordination of all design, production and erection works, scheduling determines the necessary information about the time required and duration of the various activities, deadlines and capacity requirements of erection teams and equipment.

In order to be able to effectively manage, supervise and control the sequence of work and progress from the start of a project, it is advisable to use a multi-stage planning system for larger projects. This is generally divided into general, project and detailed deadline scheduling.

After the itemisation of the project into activities and the associated cost estimation, the coordination of the project deadlines to the design work and the first works on the construction site like perhaps foundations, footing and earthworks can take place. Building on this, a project schedule with dates given for structural design, production design, ordering of materials, production, delivery of materials and erection (Fig. 153). A detailed programme of works can then be drawn up containing the exact sequence of erection. This should contain the construction site setup, provision and installation of erection equipment, preparation and preassembly work, the implementation of the erection according to components and finally the dismantling of cranes and other erection equipment and clearing the site.

The time required to complete a membrane project is generally determined by the extent and complexity of the proposed structure. From schematic design, design and achievement of building permission through the manufacture of materials and then to completion of erection, the entire process, according to information from fabricators, normally lasts an average of 6–9 months. More simple structures can be implemented more easily. If the construction project is more ambitious and the structure more complex, like for example the roofing over of a stadium, then the overall duration till handover can require 12–15 months.[2]

"… The design phase from idea to building permission normally takes 2–3 months. The time for obtaining permissions is often difficult to influence … After building permission has been obtained, than perhaps about 1–2 months for the workshop drawings and, according to the extent of the project, about 3–6 months for production at the works. The actual erection on site is often comparable to a prefabricated house. The foundations have already been completed in parallel to the works production. For the erection, e.g. for an area of approx. 500 m², about 1–2 weeks need to be reckoned with."[3]

---

1 Petzschmann, E.; Bauer, H. (1991)

2 Cenotec (1999)
3 Rudorf-Witrin, W. (1999)

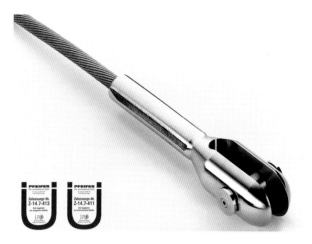

Fig. 152: Rolled fork fitting with general type approval under building regulations

An important factor to be considered in the scheduling can be the time required for project coordination, especially if many stakeholders and consultants are involved in a project. Agreements concerning contractual matters and formal decision making between client and consultants as well as alterations can have a considerable influence on the time required and in some cases can use up 50 % of the time allowed for the project implementation. If only client, architect and structural engineer are involved, then the overall duration until construction completion is normally considerably shorter.[1]

In summary, the time taken depends on factors like project coordination, the type, size and quantity of the materials used, the preparation for construction and the consideration of site practicalities in the design of cutting-out and details.

In order to achieve the economic aim of shortening the project duration, the scheduling must also take into account the time from ordering the materials to delivery.

The time from placing the order for the membrane through design to delivery on the construction site will be, depending on the number of panels and for a project of average size (approx. 2,000–3,000 m^2) and assuming clearly discussed and agreed construction methods, approx. 2–3 months. It should be pointed out for deliveries of membrane that the fabricator has normally already ordered the raw fabric before design and patterning. The quantity of raw material to be ordered is normally determined by the fabricator or the engineer responsible for patterning using values from experience.

The manufacturing time for wire ropes is about 4 weeks for common diameters and lengths, with special orders taking about 8–12 weeks. Apart from the availability of the wires, the weight and the unit weight (loading and transport), the delivery time for wire ropes can be influenced particularly by material and construction of the rope, considering the capacity of the production machinery. The capacity to produce larger weights and diameters is not available in every country. When small quantities are to be ordered, attention must be paid to the minimum production quantities for each diameter. Further factors, which can influence the delivery time, are the type and size of the end fittings. To avoid programme delays, wire ropes, anchorages and pins should be specified with general type approval, in order to avoid the need for extensive testing.[2]

---

1 Teschner, R. (2004)

2 Stauske, D. (2000)

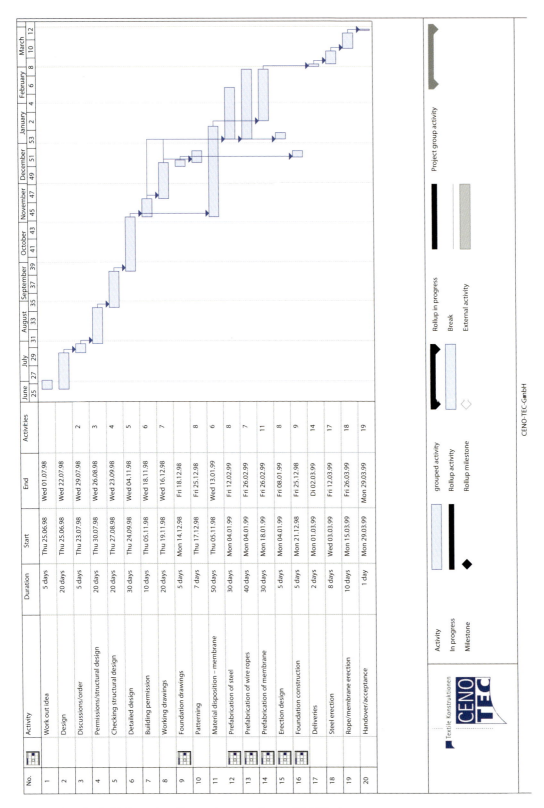

Fig. 153: Possible schedule for a project with a membrane area of approx. 500–1,000 m²

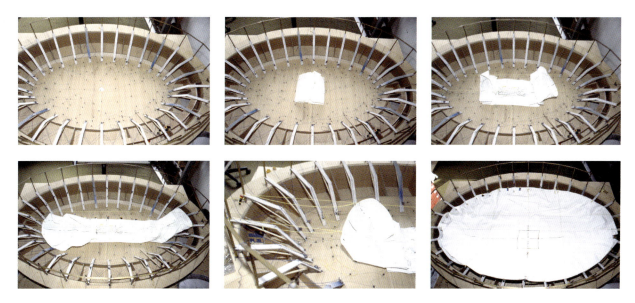

Fig. 154: Erection model for the roofing of the arena at Nîmes, France: laying out the pillow

3.2.3 Modelling erection procedures

The erection of flexible sheeting and ropes differs substantially from conventional methods of constructing structures. Procedures like laying out, lifting into place, hanging and pretensioning require the use of specialised equipment and techniques.

One aid to visualising and simulating methods of erection is the building of erection models. Especially for large projects with complicated geometry or difficult local conditions, processes like laying out, lifting and pulling of membranes can be evaluated and optimised well using models. In addition, building models can help to understand the space requirements for the temporary equipment needed and the consequences for progress of the arrangement of the equipment.

Such models are good value for money and relatively easy to build. The spatial visualisation and clear presentation enables the reduction of sources of errors at each phase of the erection. According to the design phase, models can be made in various scales. Models are especially suitable for the simulation of unfolding procedures for heavy membrane packages or for the pulling-in of complicated panels.

For the planning of the erection of the roofing of the antique arena in Nîmes, the method for the erection of the pillow structure was developed with the help of a construction model. The challenge of this construction lay in lifting an elliptical fabric pillow of 4,000 m² area complete with its supporting rope net into the anchoring position without damaging the stands in a valuable ancient monument. The weight of the upper and lower membranes to be lifted was 8 t each.

The sequence in Figure 154 shows the simulation of the unfolding process of the lower and upper parts of the pillow. Figure 155 shows the actual unfolding operation on the construction site. The erection of this project is documented in section 3.5.2.2.

Fig. 155: Laying out the lower and upper membranes for the roofing of the Arena in Nîmes, France

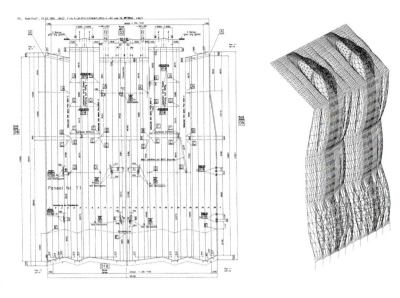

Fig. 156: Trial erection for the wrapping of the Reichstag in Berlin, Germany: Trial unfolding on a scaffolding simulating the geometry of the Reichstag façade in Konstanz, Germany

Erection models are only suitable for structural investigations to a limited extent. For the system of supporting ropes, it is possible using catenary models to determine at which points in the system hinges should be assumed, but to investigate or simulate the load-bearing capacity of individual elements or structural systems is not possible using such geometrical models. To build such a model, it is necessary to observe rules of model similarity and take the relevant stiffnesses into account.[1]

Another measure, which can be used to evaluate and test erection procedures, is the implementation of trial erections. This enables each stage of the erection to be tested, optimised and evaluated on a 1-to-1 scale under realistic conditions, making it possible to discover any mistakes in the co-ordination planning at an early stage. Trial erections are also well suited to check and, if required, make corrections to the geometry. The results of measurements undertaken on the geometry can be imported into CAD files and compared with the correct geometry (Fig. 156). Any deviations can be corrected through alterations to subsequent elements.

Above all, if the erection programme is very tight, trial erection is an important tool for construction management. It is also helpful to carry out trial erections when preparing for series production or for contracts in far-off lands.

Figure 157 shows a model in scale 1:24 for the erection of a star-shaped stadium roofing in Riyadh (Saudi-Arabia). The pulling-in of an internal ridge and valley panel of approx. 1.700 m² was simulated with an erection model in a warehouse in Buffalo. This method of erection was tried for the first time for this project.[2]

Fig. 157: Model of the erection of the roofing-over of the stands in the stadium of Riyadh, Saudi Arabia

1 To simulate the structural reality using a geometrical model, the relationship of weight to volume (8-fold increase of weight according to volume) and the relationship of the load-bearing capacity of the structural element with the increase of cross-section (factor 4) must be reconciled. This requires the application of an additional loading to the model.

2 Philipp Holzmann – construction documentation (1988)

3.2.4 Construction engineering

Economic and technical fields of activity cannot always be considered separately in construction management. The method of construction depends not only on the available finance and time allowed but, especially for the construction of wide-span lightweight structures, depends on the material properties and the engineering design and work preparation. This makes it necessary to look into the aims, tasks and areas of influence of construction engineering.

Construction engineering includes the creation of all the documents for the individual assembly units and construction components for the technical implementation of the erection procedure to be used.

This includes the coordination of design, production, transport and erection and the making of decisions about the size and geometry of erection units as well as lifting capacities, temporary construction and erection sequence.

This essentially includes the following categories:

- Advice about erection in the design process,
- Creation of an erection plan,
- Determination of the erection procedure,
- Production or organisation of all structural design checks for temporary conditions during erection,
- Checking the alignment of the cut-out strips considering the erection sequence,
- Checking the practicality of details and working drawings for erection,
- Production of all temporary works drawings and erection instructions,
- Creation of a folding and packaging scheme,
- Production of all construction component and transport lists,
- Specification of all scaffolding, erection equipment and temporary works,
- Organisation of the site set-up
- Supervision and documentation of the erection implementation.

The purpose of construction engineering measures is to make the working conditions on the construction site easier, taking into account all required conditions, and to exploit the possibilities of industrial production as much as possible.

The following schematic diagram is intended to explain the area of influence and responsibility of construction management in the design and construction process of membrane structures regarding the practicalities of handling and erecting flexible elements in the fields of design, production, transport, storage and assembly.

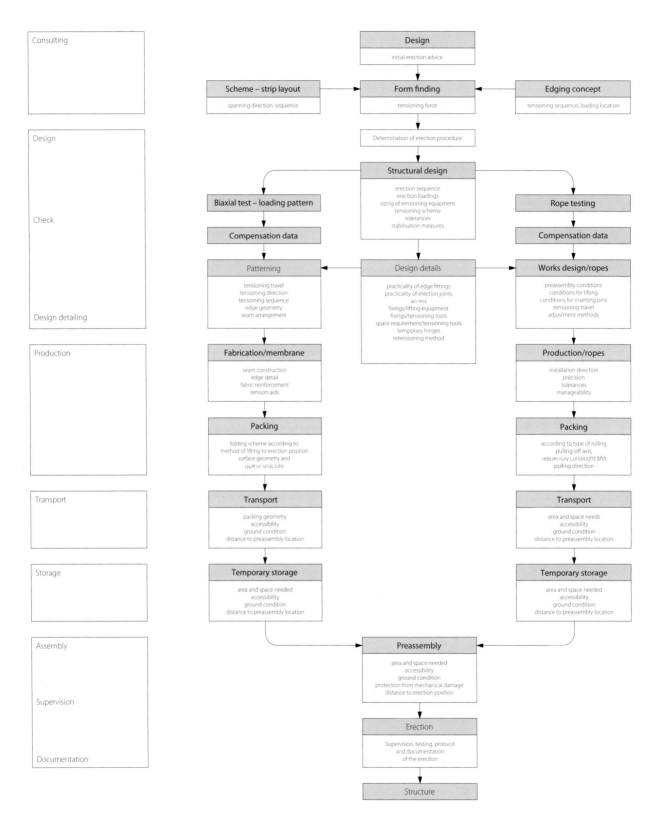

Fig. 158: Areas of influence and responsibility of construction management in the design and construction process of flexible structural elements

3.2.5 Design detailing for erection practicality

The consideration of erection practicalities in the design of details forms an important part of the technical design of membrane structures. It is a precondition for reducing the cost and duration of erection. The connections must be optimised for weight saving, manageability, precision, tolerances and possibility of lifting with the intended equipment.

The manufacture of sheeting connections is normally done by the fabricator in the works. All stages of the erection should be discussed with him. The edge connection, with the exception of webbing belts, is assembled on the site by the construction firm, except edge hem or keder seams which are prefabricated.

In addition to the manual work of preassembly, where assembly joints and membrane edges are prepared for lifting, hanging and pretensioning, the application of lifting and pretensioning devices also needs to be taken into account in the design of the connections. Loading during erection from load introduction points, which alter during the erection procedure, cannot be allowed to affect the structural function and manageability of the connection. It must be possible to alleviate deviations from tolerances through appropriate details.

Joint type/production	Erection method	Connecting element	Connection function	Location of load introduction
Welding	—	 fabric or foil	bonding	surface
Glueing	—	 fabric or foil	bonding	surface
Sewing	—	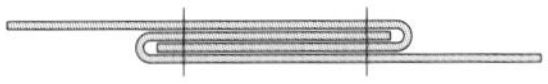 fabric or foil, seam	mechanical	surface
Welding or sewing	—	 fabric, webbing (sewing thread)	bonding or mechanical	surface or edge
Welding	pull-through	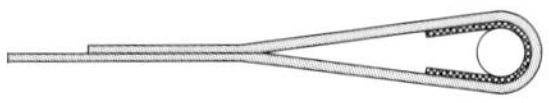 fabric, rope or tube	form-fit	edge
Welding	hanging clamping bolting	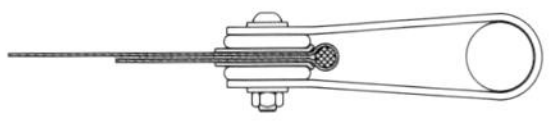 fabric or foil, keder, clamping plate, bolts, clip, rope or tube	form-fit mechanical	edge
Welding	pull-through	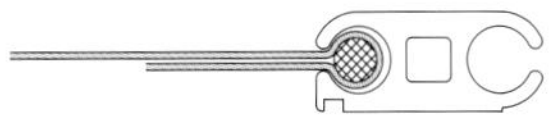 fabric or foil, keder, keder rail	form-fit	surface or edge
Welding	hanging clamping bolting	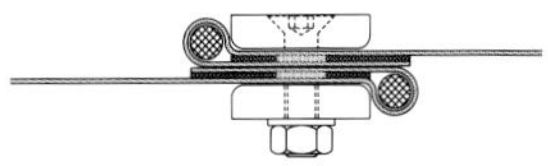 fabric or foil, keder, clamping plate, bolts	form-fit mechanical	surface or edge

Fig. 159: Manufacture of characteristic connections in membrane construction

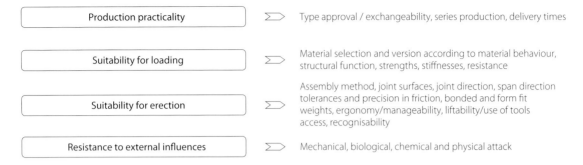

Fig. 160: Specification requirements for the detailing of membrane connections

It is important that there is adequate room for erection in the detailing of assembly joints, edges and corners. In order to reduce the risk of accidents during erection, all connection locations should be easily accessible.

The requirements resulting from loading, production and erection as well as resistance to external attack for the connections in membrane construction can be summarised as follows:

The following requirements have to be taken into account in the development of joint detailing: geometrical dimensions, weights, relevant properties during transport, erection and as part of the complete structure, functionality and material properties.

The construction company is to be handed a complete drawing list with summaries and the relevant working drawings. To explain the assembly procedures, the working drawings should contain a view of the components appropriate for erection, and there should also be a description of the constraints affecting erection. This should contain the following details:

- A clear description of the component with details of weight, dimensions and maximum permissible loads,
- Details of erection position, installation and transport location,
- Number and layout of the panels to be erected and the assembly joints,
- Details of assembly location, process and aids,
- Details of stability, maximum loads and strengths,
- Lifting points for lifting and transporting on the construction site,
- Description of the connection points for applying the pretension

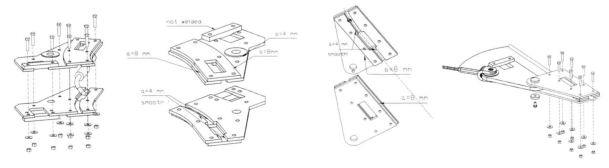

Fig. 161: Exploded drawing of two membrane corner fittings

3.3 Erection equipment and machinery

The surface and linear elements used in membrane construction can have impressive dimensions and weights. They often have to be lifted and installed at great heights on the construction site.

The lifting, lowering or moving horizontally of loads is done with various types of mechanical lifting and transport devices. For the erection of membrane structures, this normally means cranes and small lifting equipment, using appropriate load suspension devices.

Flexible linear and surface elements have to be tensioned by holding, pulling and fixing. The use of appropriate temporary construction and equipment is also required. Suitable scaffolding has to be erected and maintained in order to work with tensioning devices and working platforms often at impressive heights.

In the following section, the equipment and machinery required for the erection of membrane structures, its use and method of operation, are described.

The selection of suitable equipment for erection on the construction site normally depends on the intended erection procedure. The economic and technical criteria for the choice of equipment are explained in section 3.4.1.4.

3.3.1 Cranes and lifting devices

Cranes count as large lifting devices and are mainly used in the field of membrane construction for lifting structural elements, membrane panels and edge elements, as well as tools and temporary construction. In addition, they are often used to stabilise the primary construction (see section 3.3.4).

Any type of crane can be used, which meets the requirements for lifting capacity, reach and lifting height. Because of their very high investment cost, cranes have a considerable influence on the total erection cost and require exact determination of the necessary working time.[1] The use of cranes also influences the organisation of the erection and the working environment on the construction site. The cost of crane erection depends on the type of crane, its delivery condition and level of equipment. The types of crane most used for the erection of membrane structures are summarised in Figure 162.

Tower cranes can be used, whether with cantilever jib, luffing jib or telescopic jib. Guyed derrick cranes can also be used.

The most important characteristic of a crane is the load moment (mt). This can be calculated from the product of reach (m) and lifting capacity (t).

For the lifting of loads to great heights, tower cranes are mostly used. These are installed in a fixed position and the amount of space taken up is small. Stationary tower cranes can be built as self-erecting or jack-up or climbing cranes, which can be extended or raised. Travelling tower cranes can travel at the erection location on pneumatic tyres, crawler tracks of rails.

Tower cranes can have high or low-level turntables. The commonly used types of jib are horizontal jib with trolley and luffing jib. Both types of jib can be mounted on a turntable under the tower (tower and jib turn) or at the top of the tower (only the jib turns).[2]

Tower cranes with the turntable under the tower and trolley jib are mostly built up to a load moment of 120 mt, with luffing jib up to approx. 1,500 mt. Compared to comparable cranes with high-level turntable, they have lower drive power, a lower centre of gravity and lighter mast construction. They can be erected without a great amount of work with winches and fitted with ballast. They can convert to transport condition by folding and telescoping the jib (right in Fig. 163).[3]

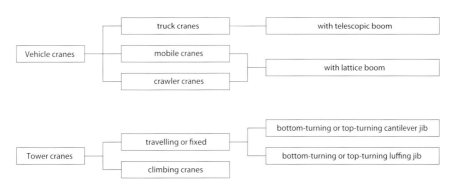

Fig. 162: Types of crane frequently used for membrane construction

---

1   Petzschmann, E.; Bauer, H. (1991)

2   DIN 536-1 to DIN 15030 (1995)

3   Drees, G.; Krauß, S. (2002)

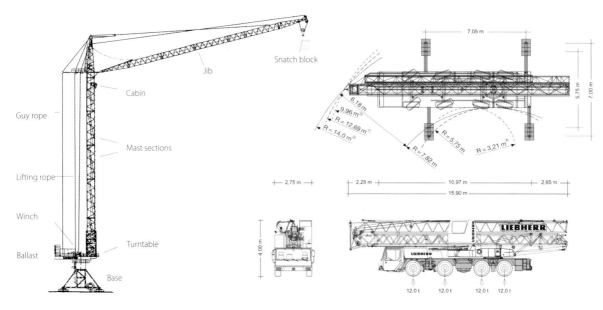

Fig. 163: *left:* Bottom-turning tower crane with luffing jib; *right:* Transport condition of a Liebherr mobile tower crane

Tower cranes with high-level turntable mostly have a nominal load range of up to 120 mt. The ballast and the lifting winch are situated at the end of the opposing part of the jib, an advantage where space is restricted and there is no room for turning gear and ballast on the ground. When they are equipped with a jack-up mechanism, they can be extended higher without having to use another crane (right in Fig. 164).

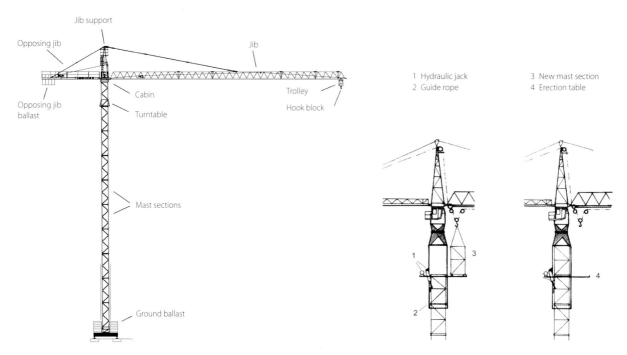

Fig. 164: *left:* Tower crane with trolley jib and high-level turntable; *right:* Climbing system for a tower crane with high-level turntable

Truck crane with telescopic boom

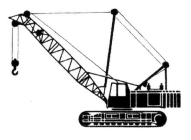
Crawler crane with lattice boom

Mobile crane with lattice boom

Fig. 165: Mobile cranes

Cranes with wheeled chassis are available as road-going versions, with telescopic or lattice jibs. Truck cranes are available for loads up to approx. 1,200 t and are mostly hired for erection work.

Mobile cranes on pneumatic or solid rubber tyres only have a restricted speed and are therefore mostly used within a small radius. The main turnover of new cranes today is with rough terrain mobile cranes. Such machines have hydraulic suspension and are steered on many axles.

Crawler cranes are similarly built to dragline excavators. They can lift loads and transport them on the construction site. Crawler cranes are available for lifting loads of 300–1,200 t. They are most capable of travelling off-road and are used above all where steep gradients have to be climbed. Crawler cranes are useful for the installation of heavy elements. Their low ground pressure makes them especially useful for use where the terrain is unfavourable.

Truck cranes are mounted on a truck chassis. They can mostly be erected without assistance and are therefore very flexible. This flexibility makes the truck crane the most-used large lifting device in steel and membrane erection. It is preferred for short and quickly changing erection applications.[1]

The truck crane is specially built for travelling on asphalt roads, reaching travelling speeds up to over 60 kmh. The chassis mostly has 2 to 9 axles with the front and rear axles steering. Larger cranes have separate operator cabins for travelling or working.

Truck cranes are available with telescopic or lattice boom.

The major advantage of the telescopic boom is the possibility of altering the boom length quickly without dismantling. Under certain conditions, it can be extended or retracted under load. The telescopic boom consists of rectangular or oval section steel profiles with good buckling resistance, which are powered by a hydraulic cylinder. The reach and lifting height of a truck crane can be extended by elongating the boom with a lattice jib, making a hook height of up to 150 m possible.[2]

Lattice booms can master great lifting heights and reaches with a low dead weight, and are especially suitable for long-term applications, where the boom length does not need to be altered.

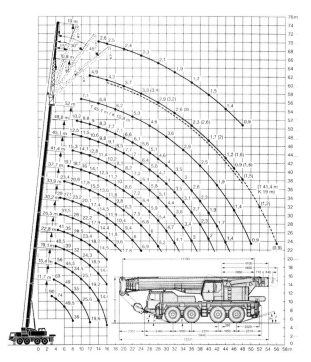

Fig. 166: Liebherr truck crane in working and driving configuration

---

1 Petzschmann, E.; Bauer, H. (1991)

2 Drees, G.; Krauß, S. (2002)

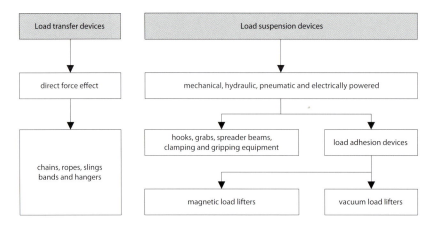

Fig. 167: Load transfer devices and load suspension devices according to power

When working with truck cranes, permissible wind speeds have to be observed.

Cranes and lifting devices must be equipped with appropriate load suspension or load transfer devices to pick up or attach loads. These are located between the load and the crane hook and do not belong to the lifting device.[1]

As shown in Figure 167, load suspension devices can be differentiated from load transfer devices by the method of force applied to the load.

Load transfer devices must not produce deformation under loading, and there should be no change of position when the load is set down. Elements, which have to be turned from vertical to horizontal or vice versa, have to be prevented from slipping.

When lifting with load transfer devices, these are laid round the load and hung on the lifting device directly or with an end connection. End fittings are available in many types, strengths and versions.

Fig. 168: Load suspension and load transfer devices for lifting a rolled membrane

1 Ludewig, S. (1974)

3.3.2 Tensioning devices and equipment for ropes

Appropriate devices and tools have to be used to apply the forces to the edge, carrier, tension and stay ropes of membrane structures. These are key devices on the membrane construction site. In the construction industry, these are mostly known as *lifting devices loaded in tension*. Because the tensioning of load-bearing elements is a central process in this form of construction, these are called *tensioning devices* in this book.

To deflect, guide and apply the tension forces, it is often necessary in addition to make use of suitable devices to secure the location of the tensioning device relative to the load-bearing element. Such tensioning mechanisms in combination with the tensioning devices form tension systems, which can be electrically, hydraulically or mechanically driven.

3.3.2.1 Electrical tensioning systems

Rope winches, block and tackle

Rope winches have many uses. They can be used to control movement and as tensioning devices for heavy load-bearing cables. Manual and motorised rope winches are available, and they can be equipped with a wide range of accessories. As tensioning devices, rope winches can also be used to operate through tackles. They always need to be anchored and protected from overloading.

One of the oldest methods of lifting and lowering loads is working with block and tackle. The principle of the block and tackle is based on the loading being distributed by the arrangement of pulleys onto many runs of rope.

For tensioning ropes, tackles are mostly used when long tensioning travel distances are needed. For very long rope lengths, they are mostly used in combination with rope winches, a procedure also used in the construction of cableways. Tackles with 2, 4, 6 or more pulleys can be used according to the force to be applied. The parallel arrangement of many runs of rope means that the load applied is many times the load needed to pull the tackle. The rope runs alternatively over the pulleys of a fixed and a movable block. The main types are the long thin rope puller with pulleys of different diameters and the wide, short form with pulleys of the same diameter.[1]

Tackles can be used for loads up to approx. 1,200 kN, their assembly is time-consuming, but they have many purposes and can be used many times.

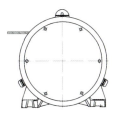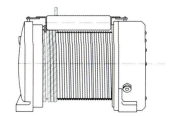

Fig. 169: Electric rope winch

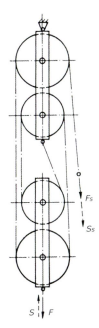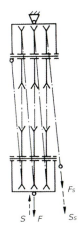

Fig. 170: *left:* Longer and shorter advantage tackle with fixed and loose pulleys; *right:* 600 kN tackle

---

1 Scheffler, M. (1994)

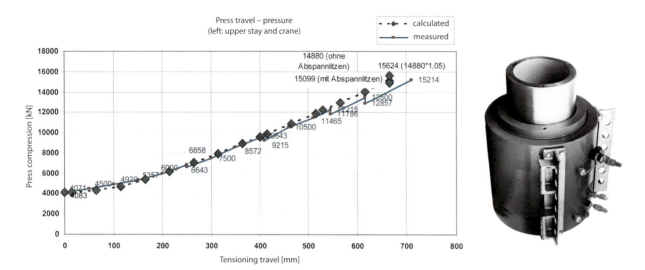

Fig. 171: *left:* Press graph; *right:* Hollow piston press

3.3.2.2 Hydraulic tensioning systems

Hydraulic jacks

The most commonly used tensioning devices are hydraulic tension and compression presses. They consist of two cylinders of different cross-sectional areas connected to each other, which are both closed by a piston. Coupled as a system with a liquid cylinder and a pump, they act on the principle of confined fluids. Enlargement of the piston area leads to an increase of the force acting on the piston. Tensioning with hydraulic presses is very precise; with press tensioning equipment, a thread play of only a few millimetres is practical. Another advantage of tensioning with presses is that they can be coupled to each other and individual press circuits can be tensioned under central control and monitoring.

Tension and compression presses are available from 100 kN up to about 7,000 kN tensioning force. Up to about 7,000 kN and corresponding rope diameters, hollow piston cylinders can be used; when the loads are higher, devices of a larger build are used. The pressures to be exerted by the press are given for each loading case during erection. The effective force in the wire rope can be read and recorded from calibrated manometers.

Tensioning equipment for hydraulic presses

In order to be able to tension single wire ropes and groups of wire ropes with hydraulic presses, temporary construction like press tables or tensioning equipment has to be used, which make it possible to guide the tensioning forces safely.

Fig. 172: *left:* Tensioning set-up for a stay rope; *right:* Tensioning set-up for a group of ropes

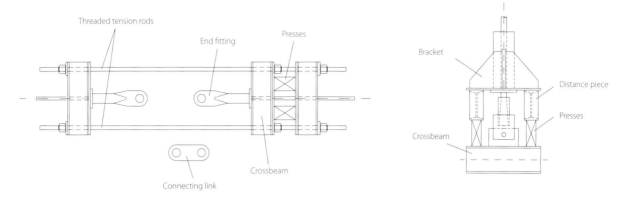

Fig. 173: Equipment for tensioning a free rope length *(left)* and at a foundation *(right)* with hydraulic presses

These mechanisms are normally designed by the construction firm and constructed according to structural requirements. They mostly consist of threaded tension rods, which are pressed through backplates by presses. For tensioning at locations with varying rope gradients, hinged tables adjustable for height and length can be used with press chairs behind them.

The technical effort to manufacture a tensioning mechanism and the amount of time and work to assemble it can be very considerable. The assembly and hanging of a tensioning system for a rope bundle can last many days, but the tensioning itself only lasts a few hours. If it is assumed that the amount of time and work to assemble a tensioning system is often many times that of the actual tensioning process, then it is clearly advantageous to tension with as few presses as possible. Attention needs to be paid to the space required for such a setup and its heavy weight. They often weigh many tonnes and can only be moved by a crane. In additional, such equipment is often an expensive one-off product, and can normally only be used once.

If tensioning is to be done in mid-air, then the presses work against a system of 3 crossbeams (left in Fig. 173). After applying the tension, the ends can be finally connected with a link.[1]

The illustration at the right in Figure 173 shows the tensioning equipment for a stay rope. If the wire rope is pretensioned with compression presses, then the presses are supported at one end on the tension member to be tensioned and at the other on a crossbeam, which is connected to the final connection with a nut.

The picture in Figure 174 shows the setup to pretension a stay rope for a pylon of the membrane roofing in Bad Velbert.

Fig. 174: Equipment for tensioning a stay rope

---

1 Ferjencik, P.; Tochacek, M. (1975)

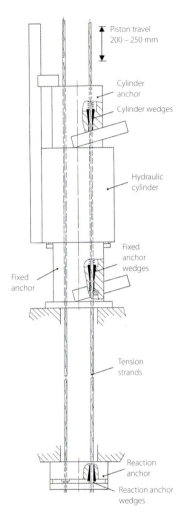

Fig. 175: Strand tensioning system

Strand tensioning devices

To tension higher loads, strand tensioning devices are often used. They consist of hydraulic jacks, steel strands, cylinder anchor, fixed anchor, reaction anchor and locking wedges (Fig. 175). They are simple to use, working similarly to a griphoist (tirfor). The individual strands are pulled by locking wedges, which close hydraulically or mechanically and connected with a clamping plate.

Strand tensioning devices can be used for up to 7,000 kN tensioning force. The maximum tension force applied per rope for the VSL strand tensioning device shown left in Figure 176 was 3,800 kN. The highest tensioning force ever installed was produced by 88 devices for 44 wire ropes with a total tension force of 214,400 kN.[1]

Long pulling travel distances (5 – approx. 200 m strand length) are possible with this system. These lengths are, however, dependant on the type of structure and the way the ropes are laid out. The puling travel is normally about 25 m. The strand lengths are, according to where the device is mounted, about 5 – 20 m longer.

One disadvantage of this system is that it is only possible to pull relatively slowly. According to the forces applied and the size of the device, pulling rates of 3 – approx. 30 m/h can be achieved. For the last tensioning distance, where the tension force is many times higher, rates of only 1 – 2 m/h can be achieved. To pull 25 m of wire rope may require 3 days.[2] During the tensioning period, various measurements have to be recorded and intermediate calculations carried out.

One single rope is normally pulled by 1 or 2 devices, seldom by 4. Up to 20 devices can be controlled simultaneously.

Fig. 176: VSL strand tensioning device and the application for the installation of the cable stays in the construction of the Waldstadion in Frankfurt/M., Germany

1 Junker, D. (2004)
2 Inauen, B. (2003)

Fig. 177: *left:* Installation of the rope clamp and the deflection system; *right:* Installation of the guide traverse and hanging the rope

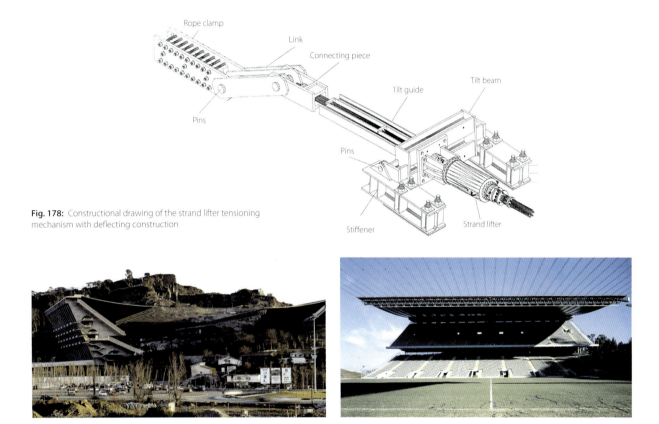

Fig. 178: Constructional drawing of the strand lifter tensioning mechanism with deflecting construction

Fig. 179: Football stadium at Braga, Portugal after the roof erection

Constructional measures are necessary to install a strand lifter. For the rope end, holding pieces or traverses, and for the tension device, support brackets or rockers. The technical work involved in installing a strand tensioning device depends on the location of the tension anchor and the length of the strands.[1] It is important to arrange the tensioning devices exactly in the axis of the rope to be tensioned.

In order to also be able to tension with strand tensioning devices at an angle to the rope axis, clamps with deflection mechanisms are used. These ensure tensioning of the rope to the anchorage without damage. Figure 177 shows the preparation work for tensioning the carrier rope of the hanging roof construction of the stadium in Braga (Portugal) in July 2002.

1 Eberspächer company information (2003)

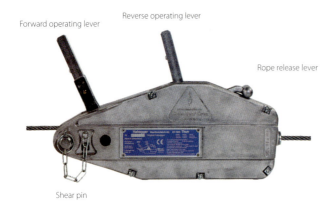

Fig. 180: *left:* Mechanical griphoist; *right:* Tensioning a rope with a griphoist

3.3.2.3 Mechanical tensioning systems

Griphoist and ratchet lever hoist systems

Griphoists (tirfors) are manually operated winches, which are mainly used for pulling, tensioning, lifting and lashing loads without limit to the rope length. They work on the principle of an eccentric clamp grip. A wire rope of any length is drawn through the device using gripping jaws. The pairs of jaws are operated by pulling a lever backwards and forwards. The changeover from pulling to releasing can be done by changing the location of the operating lever, even under load. The load in this case is held safe by automatic closing of both pairs of jaws. A shear pin safety device on the feed lever protects from overloading. By using a deflecting block, the load can be distributed over many ropes, allowing a doubling of the tension loading.

Griphoists can be used for tension loads of 3 – 100 kN during erection work. They have a low dead load and the operation is simple.

Where space is tight, loads can also be moved using ratchet lever hoists. They are used for loads up to approx. 100 kN and are operated by moving a lever backwards and forwards. The movement direction can be changed by changing the position of a lever on the ratchet arm.

Both tensioning devices need care to be taken during tensioning, because they can relatively easily be overloaded and damaged by too much strength. Fabrics and ropes can also be loaded to breaking load and damaged by incorrect use. It is therefore important to read the tension force from a measuring device during tensioning and observe the stated loads for the devices.

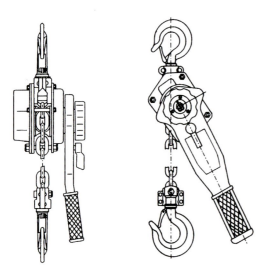
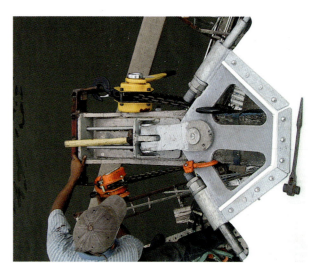

Fig. 181: *left:* Ratchet lever hoist; *right:* Moving a corner fitting using a lever hoist

Ratchet lever hoists can also apply forces to single ropes. Figure 182 shows the tensioning of a stay rope for a high point membrane structure in Zeltweg (Austria). The alignment of the pylons for the delayed application of point loads into the stiff membrane was done there using lever hoists. To shorten the rope, belts, bands and butterfly plates were used between rope and tensioning device.[1]

Wire rope grippers

Rope grippers are a simple-to-use tensioning tool for ropes. In addition to versions as eccentric clamp, rope grip and wedge-lock grips, a variety of special forms are available. Rope grippers are mostly made of hardened wrought iron with tension springs. The components are riveted to each other and pivoted. Because of their light weight (up to about 0.1 kN, they are easily manageable.

In use, a tensioning device working on the ring operates the clamping jaws, which pull the rope through their grooved gripping channel. Wire rope grippers are self-gripping, meaning that increasing tension grips the rope tighter and increases the holding strength. Because of the effect of the spring, rope grips still hold the rope fast without tension.

Turnbuckles

When the necessary load to tension the rope can be brought by hand, then ropes can be shortened with turnbuckles. This is done by turning the nuts on the turnbuckle, which act on opposing threads. Turnbuckles are normally situated near the end connection of the rope.

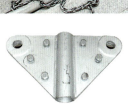

Fig. 182: Tensioning a stay rope with ratchet lever hoists and butterfly plates

Lock grip   Wire rope gripper   Eccentric clamp rope grip

Fig. 183: Wire rope grips

Fig. 184: *left:* Forked turnbuckle; *right:* Turnbuckle with eye

---

1 Lenk, S. (2004)

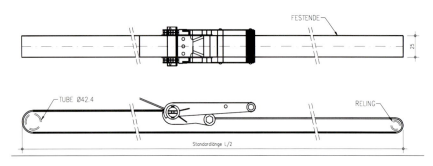

Fig. 185: *left:* Lashing strap system; *right:* Ratchet tie-down

3.3.3 Tensioning devices and aids for membrane sheets

To apply the force to a membrane sheet, it must be held, pulled with a defined force and fixed. Various tools are used for this process, depending on the material, the tensioning travel and the type of edging. These are normally electrical, hydraulic or mechanical tensioning devices and tensioning aids gripping the membrane edge.

Examples of the most important tools for pulling in and tensioning membrane sheets are described in the following section and their function explained.

Stiff membrane edges can be easily tensioned using lashing strap systems, such as are mostly used as tie-downs for securing loads (Fig. 185). Low-stretch, woven polyester straps equipped with lever ratchet tensioners are a tensioning system, which can apply permissible loading of up to approx. 200 kN. The advantage of tensioning with lashing straps is their manageability and low weight. Tie-downs can easily be installed in restricted spaces. As with all tensioning devices, they do need adequate anchorage (Fig. 186).

If higher forces need to be applied to the membrane edge, hydraulic presses can be used. If the tensioning involves long travel distances, then rope winches can also be used.

Fig. 186: Tensioning a stiff membrane edge using a lashing strap system

Fig. 187:
left: Tensioning with rope winch;
right: Tensioning with hydraulic press

In order to move membrane sheets to their intended anchorage position, they are can be lifted or slid along linear load-bearing elements. Once there, they have to be tensioned and fixed in accordance with the structural design. This is often done by arranging tensioning aids between the tensioning devices described above and the membrane edge to enable safe introduction of tensioning force into the flexible sheeting.

Tensioning aids are mostly formed steel parts designed to grip either the keder or an entire fitting. The important design requirements are the configuration of the edge detail, the extent and type of the tensioning force to be applied, the tensioning direction and the tensioning travel. The amount of space available and the type of fixing of the tensioning aid to the substructure are also relevant in the selection of equipment.

One tool used by many erection companies is the temporary clamping plate. This can be used to pull in the membrane sheet and also for tensioning. It has lower and upper clamping plate halves and connections for flat steel, ropes or belts as required. According to the exact version, the keder may lie between the plates or at the outer edge of the plates (Fig. 188).

Temporary clamping plates are used to grip the keder at defined spacings, with the edge being pulled over the anchoring point. The free keder edge is then laid in the lower half of the clamping plate and the upper part is bolted on top.

If the membrane is to be pulled along rails mounted on curved members, then roller trolleys can be mounted on the temporary clamping plates (Fig. 189). The trolleys are then pulled with tirfors. When pulling membranes along keder rails mounted on curved members, the high friction between keder and rail can be a problem.

When stiff clamping plate edge details with external keders are specified, sheet metal plates with welded-on lugs can be used for tensioning (Fig. 191). For the tensioning process, these are bolted directly to the upper part of the clamping plate. A tensioning device can be hooked up to the hole in the lug, which provides the necessary force to the edge. The drilled hole thus has to be sufficiently large.

Fig. 188: Temporary clamps for pulling in membranes

Fig. 189: Pulling in the membrane for the roofing over of the Gottlieb Daimler Stadion in Stuttgart, Germany

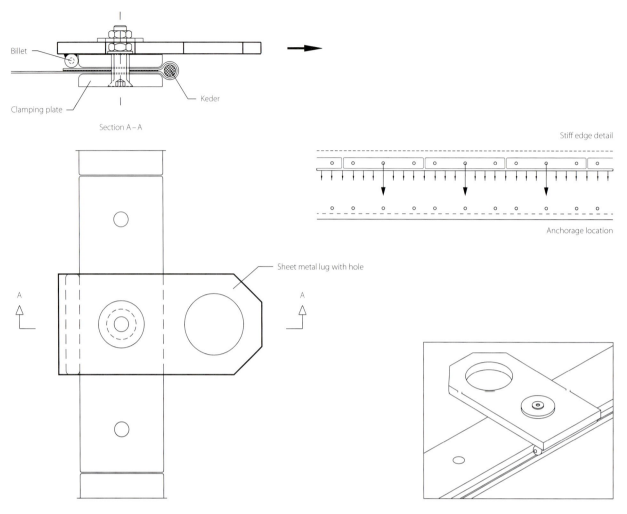

Fig. 190: Tensioning equipment for clamping plate edge

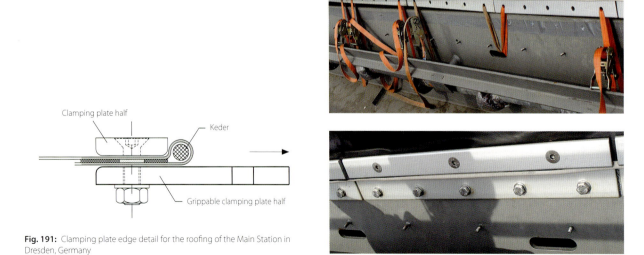

Fig. 191: Clamping plate edge detail for the roofing of the Main Station in Dresden, Germany

Fig. 192: Longitudinal pulling of a keder rail edge

Fig. 193: Tensioning a keder rail edge at the control tower in Vienna airport, Austria

The force transfer during the tensioning process is through bolts arranged in a line onto the keder. On reaching the anchorage position, the clamping plates are bolted to the substructure and the plates removed.

If it is conceivable from the design point of view and there is enough space, it can be advantageous for the erection process to alter the dimensions of the lower clamping plate half. The detail shown in Figure 191 can be gripped directly with the tensioning tool, pulled and bolted to the substructure.

If there is not enough space to use temporary clamping rails, then the keder can be gripped with a specialised tool as shown left in Fig. 192. This tool is also especially suitable for longitudinal pulling of membrane edges, where the pulling is through welded-on tension lugs with prefabricated keder.

Stiff edges with keder profiles can be tensioned with tools, which grip the keder profile directly. The edge can be pulled to the anchorage position with a lashing strap system.

Figure 194 shows how much potential for optimisation can lie in the detailing of the edge for ease of erection. Where space is limited, the provision of temporary metal plates as anchorage for tensioning tools can be extraordinarily helpful (left in Fig. 194). After the tensioning is complete, these can be dismantled and reused. Steel loops welded to the primary structure are also suitable for anchoring (right in Fig. 194).

Fig. 194: Equipment for fixing tensioning devices and tensioning aids

Fig. 195: Use of system scaffolding as working platforms for the installation of membrane surfaces

3.3.4 Scaffolding working platforms and temporary construction

Scaffolding working platforms are often used to ensure safe working for the erection of membranes. Scaffolds can be stationary or wheeled depending on the structural system and the erection procedure. These often have to carry, in addition to the erectors and their tools, the membrane material, which is laid out in preparation for installation.

3.3.4.1 Stationary and wheeled scaffolds and working platforms

The use of lightweight system scaffolding is economical where the erection of the membrane is laid out axially. In order to encourage economical erection progress, there should be enough space available. Such scaffolding is normally stationary or wheeled frame or tube scaffolding with working platforms of steel or aluminium panels. The advantage of the use of scaffolding systems is that they can be relatively easily rebuilt and extended on account of their light weight and the welded or coupled connections are standardised through prefabrication.[1]

Where the erection of conventional scaffolding is hardly practicable or uneconomical, mobile working platforms are often used with scissor-lift, articulated or articulated and telescoping mechanisms. With lifting capacity of approx. 500 kg, they enable working heights of up to approx. 35 m.

The working platforms are specialist devices mounted on a truck chassis or on crawler or wheeled chassis, and can often be operated from the working platform. Machine types with special chassis can cope with terrain gradients of up to about 25 %. The advantage of using articulated, telescoping and scissors platforms lies above all in the extreme flexibility in restricted working areas.

When choosing suitable working platforms, safety and ergonomic working conditions during the erection have to

Fig. 196: Articulated and telescoping platform in use for the erection of a pneumatic pillow and the erection of a four-point awning

1   Jeromin, W. (2003)

Fig. 197: Hanging scaffolding for the erection of the Millennium Dome in London, England

Fig. 198: Travelling platforms for the membrane erection of the Tropical Island in Brand and for the Forum roof in the Sony Center in Berlin

be taken into account. Total weight, dimensions of devices, bracing widths, working heights, load capacity of the working basket, turning and slewing areas and sideways reach are the most important criteria.

Erection work at great height can also be performed using hanging scaffolds. These are hung with inflammable carrier structures to elements of the structure with sufficient load-bearing capacity and have to be secured against swinging. The structural safety of hanging construction is often ensured by ballasting.

For membrane erection, hanging scaffolds are often built to travel. Their position can be moved along guide elements like ropes or edge beams. Vertical and horizontal movements have to be secured by suitable fixing, stop or catch mechanisms. If they are fixed, then they will have to be transported by crane.

Figure 199 shows the erection of the membrane to a galet at the site of the EXPO.02 in Neuchâtel, Switzerland. To erect the membrane in the collar areas, a travelling working scaffold with lower and upper platform was developed to mount on the upper tubular edge beam. In order to achieve a stable working platform, a cantilever ballast beam was mounted on the upper platform, which pressed the lower platform against the lower part of the box-shaped edge beam. The travelling and adjustment process to adapt to the curve radius was performed manually.[1] The pillow erection of the galets will be described in section 3.4.2.

Fig. 199: Mobile working platform for membrane erection at Neuchâtel, Switzerland

1 Imgrüth, H. (2002)

Fig. 200: Abseil work for membrane erection

Scaffolds require a not inconsiderable length of time to erect and mount to fixed construction elements. The use of scaffolding or cranes is often impossible on account of the geometry of the structure, the space available or unavailability.

For erection work, which cannot be carried out with conventional scaffolding, mobile working platforms or travelling scaffolds, one possibility is working supported by ropes.

The advantages of abseil working are high flexibility and low cost. Especially for projects with wire rope systems, which can be used as access routes, abseiling is a very efficient solution. To ensure safe working with ropes, it is, however, necessary that the erectors are fully trained in abseiling.

3.3.4.2 Temporary construction for the stabilisation of the main structure

Flexible structural elements are not sufficiently stable before they have been tensioned. Structures, in which the flexible and the rigid load-bearing elements stabilise each other in the complete structure, have to be held in equilibrium for the installation of the membrane sheeting (see section 3.4.1.2).

Various methods and temporary works can be used to stabilise the primary structure or the elements of the substructure during erection depending on the erection process. Structural components can be tensioned, stayed, or propped with equipment like wire ropes, lashing straps or adjustable props to enable the erection to proceed safely. There must be adequate anchorages available for their use.

Fig. 201: *left:* Stabilising the primary structure during the erection of the roof over the carport at the Munich Waste Management Office; *right:* Stabilising the wire rope construction at the Millennium Dome in London, England

Fig. 202: Temporary state during the erection of a high point roof

One low-cost method of achieving a stable temporary state is the use of tie-downs to stay or tie the primary structural elements. Because the stability conditions alter during erection, good access to the ratchet levers is important. These need to be tightened or loosed according to the control and adjustment plan.

If construction elements are lifted by crane, then the structural stability of the devices in temporary states during erection needs to be watched (Fig. 202).

In order to maintain the structural system in equilibrium, it is often necessary to preload individual structural elements. This is the case, for example, when membrane panels or wire ropes first reach their intended geometry after the installation of stiffer elements (metal or glass, etc.). The equilibrium state of the structural system can be brought about by the application of this loading in advance. This is done by ballasting the structure in the temporary erection state to stabilise it.

Fig. 203: *top:* Ballasting a wire rope construction for glass panels with sandbags at the nodes; *bottom:* Ballasting two carrier ropes with water containers during the erection of the roof over the Gerhard Hanappi Stadion in Vienna, Austria

Preloading of a structural system can be done in various ways. Common types of ballast are water or sandbags (Fig. 203). The advantage of this process is that it is cheap and environmentally friendly. The containers can be tapped to reduce the weight. The sand running out can be reused on the construction site, the water can soak away. Water tanks can also be topped up if a water feed has been installed, which can be helpful according to the temporary state.

3.3.4.3 Temporary construction for devices used during erection

If heavy tools have to be installed at great heights, then load-bearing scaffolding will be needed. When tensioning wire ropes, the tensioning equipment has to be set up for applying the tension. Tensioning chair, press chair and the required presses have to be positioned in the rope axis and anchored. This can be provided by the erection of accessible erection platforms.

Fig. 204: Platforms for the strand tensioning devices during the construction of the roof over the Waldstadion in Frankfurt/M., Germany

3.4 Erection procedure

The sequence of activities for the erection and assembly of prefabricated construction elements and units to form a structure can collectively be described as the erection procedure.

The erection process always underlies an erection principle. In the field of wide-span lightweight structures, this principle always arises primarily for practical reasons, the purpose of which is to complete a tensioned structure.

This section is dedicated to the factors influencing the erection procedure and investigates a variety of construction processes for the erection of membrane structures.

3.4.1 Criteria affecting the erection procedure

Considering the complex tasks in the field of production, delivery and assembly, with ever more new, contrasting quality requirements, the creation of erection schemes places the highest demands on designers and erectors. Out of the wide range of possible ways of spanning large areas, a multitude of special constructions are born in the field of lightweight structures, and this results in a wide range of construction options.

The parameters, on which each erection principle is based, derive essentially from production technology, construction technology and economics. The categorisation of these factors is according to material, construction details, how loads are resisted, temporary equipment and logistics parameters.

The principle to be used for construction is thus essentially based on:

- the type and jointing of the materials to be used,
- the working principle of the structural system,
- the local conditions on the construction site,
- the type of construction equipment used and
- the production, delivery and transport practicalities.

3.4.1.1 Type and jointing of the materials to be used

The production and jointing, the behaviour of the materials used under loading, time and temperature are all of importance in the erection of membrane surfaces. This section describes how the influence of the chosen material, the patterning and jointing methods affect the individual erection procedures and activities.

When working with foils and fabrics, it is important to consider the risk of damage from production and erection activities. The way in which flexible sheeting is laid out, lifted and fixed determines how it should be packed and delivered. It is also important when working with ropes to take measures to avoid damage to the ropes.

The effect of the material on the particular erection method is most clearly shown in the introduction of tension into the flexible load-bearing element. Tensioning direction and sequence depend to a decisive degree on the material type and the method of joining the panels. The arrangement of the strip layout, the pattern and details of the seams, the form of edging and the method of support are decisive for the erection method to be used. The detailed design of the constructional elements in the areas where forces are introduced into the membrane surface under consideration of erection practicalities should also be mentioned.

The strength of the tensioning force to be introduced, the extension and the tensioning steps to be applied over a certain period of time also depend on the material. Material and processing characteristics vary widely among materials. The limitations on temperature for working with certain fabrics and foils also need to be observed.

3.4.1.2 The working principle of the structural system

To describe the criteria for selecting a suitable erection procedure, it first has to be considered what type of wide-span lightweight structure is to be constructed.

The structural form of structures with flexible load-bearing elements held in place by fixed edges or corner fittings ideally corresponds exactly to the shape of the forces. Deformation is caused when flexible load-bearing elements are placed under loading; the resulting stresses are largely tension stresses. In order to relax the peaks of stress through the deformation of the flexible load-bearing element, the type of material, the detail of the edge construction and the form of the structure are all important for the load-bearing behaviour. Structural systems with flexible elements, like membrane structures and rope nets, are therefore described as form-active structures.

Structural types according to form, support and edging

The geometry of the structure has become established as the criterion for categorisation. This gives information about the shape of a structural element. Mechanically tensioned, anticlastically curved constructions are divided into:

- *Awnings* – hyperbolic paraboloid-shaped surfaces tensioned from two or more high points,
- *High point* – surfaces pulled to one or more high points from inside or outside,
- *Arches* – saddle-shaped surfaces spanned between arch-shaped edge beams and
- *Ridge and valley* – wave-shaped surfaces spanned between high-level, linear elements.

Synclastically curved constructions tensioned by pressure differences are divided into:

- *Element pillows* – tube-shaped inflated pillows arranged next to each other,
- *Large pillows* – large-scale lenticular inflated pillows and
- *Air-supported halls* – large-format domed inflated halls.

A further wide variety of mixed forms and derivations can be derived from the basic forms listed. Structural special shapes, whose place in this categorisation of types is not clear, can nonetheless normally be shown to be based on one of these basic forms.

Membrane elements can also be used on framed substructures as room-forming covering surfaces for almost any shape.

The equilibrium shape under the action of loading plays an important role in form-active structural systems, which explains why the flexible surface load-bearing elements of these structural systems are normally divided into three groups according to their type of support:

- *Membrane surfaces supported at a point* – mechanically tensioned high point and awning surfaces, whose edges are anchored to the primary construction from inside or outside at points,
- *Membrane surfaces supported along a line* – mechanically tensioned arch and wave surfaces, which are edged by linear edge elements,
- *Membrane surfaces supported over the whole area* – pneumatically tensioned pillow constructions and air-supported halls, whose surfaces are supported by the increasing or reduction of pressure. [1]

---

1 Remark: Weighted hanging roofs made of fabric are practically never built, or else they would also appear in this group.

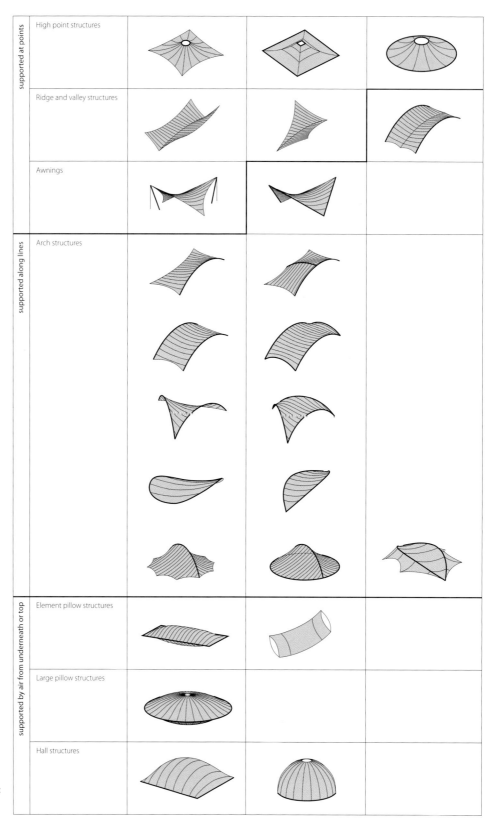

Fig. 205: Typology of structures: basic types according to their shape and type of support

Stability conditions in temporary erection states

The typology of structures described above and illustrated in Figure 205 can, however, only serve for basic consideration of the question of the choice of a suitable erection procedure. For the erection of a membrane structure, it is also necessary to evaluate the strength and stability behaviour in the structural system and the type of interaction of primary and secondary structural elements.

Flexible load-bearing elements do not have a stable form without pretensioning and thus have insufficient load-bearing capacity. This instability in the erection state differentiates the erection of form-active structures considerably in comparison to conventional structures.

Dependant on the structural system, large deformations and oscillations can be caused by loading during erection. The loading on the construction elements is then mostly higher than in the final completed state. Stability and structural safety of the entire structure in the erection state are endangered. The instability of the overall structure or individual structural components during erection is to be feared above all when flexible and stiff structural elements stabilise each other in the completed state.

The following two categories are therefore decisive for the erection situation

1. *Structures, whose primary load-bearing structure has sufficient stability without the flexible load-bearing elements,* and
2. *Structures, which only have sufficient stability through the interaction of stiff and flexible elements.*

Category 1:

With these structures, the erection of the membrane is first carried out after the completion of the primary structure, which is stable on its own. Any sub-structures can be either flexible or stiff, and may have to be stabilised before the installation of the membrane surface, be designed for erec-

a) Installation in a stable umbrella structure

b) Pillow installation

c) Installation of arch membrane

d) Installation of low point membrane

Fig. 206: Membrane installation in stabilised primary construction

tion-state loadings and may have to be articulated. The stabilisation work can then usually be done using ropes, belts or lightweight devices. If all the sub-structure parts of the primary structure are also form-stable, then the membrane with small spans is a room-forming cladding surface.

The lifting, hanging and pretensioning of the membrane are the essential erection processes. Temporary membranes and nets can be fixed to the primary structure, if the load-bearing capacity is sufficient, to provide anchorage locations for tensioning devices and equipment.

Category 2:

When erecting structures whose structural function is based on the interactive stabilisation of the primary structure and the membrane, temporary stabilisation measures also have to be provided for the primary structure. The loads acting on the structure are only in equilibrium in the final state after the intended geometry has been created. In order to avoid overloading at any temporary state during erection, the structure must be supported, held or anchored at all essential points until the structure is complete.

If the stiff load-bearing elements of the primary structure are moved to their intended final position at the same time as the membrane (Fig. 207), then the elements of the primary construction also have to be designed to resist loadings at each state during erection. The resulting deformations also have to be compensated by various technical measures, so that the components have sufficient stability in each state of geometry. It is important to note here that the loadings during erection mostly have a quite different nature and extent than the loadings in the completed state.

Temporary staying or ballasting measures can stabilise structures in temporary states. Adequate anchorage points must be available for staying with tie-downs or ropes. This can be done with additional holes in fixed parts of the structure or the provision of temporary foundations.

a) Erection of a high point structure

b) Erection of a awning

c) Erection of an arch membrane

d) Erection of high point structures

Fig. 207: Membrane erection with unstable primary structure

In order to position construction elements in their intended location, they are often provided with temporary hinges. In this case, the additional space required for the element to be hinged has to be considered. Because of the mostly rather high cost of making such hinges, it is a good idea to investigate whether an alternative erection process could not be used in such cases.[1] Costs can be reduced by reducing the number of such temporary hinges and the limitation of the degree of freedom. Special attention is necessary with the provision of ball-jointed hinges. The tensioning force must be in just the right height and direction to hold the ball on the bearing during tensioning.

The weather conditions during the erection are also an important aspect. Strong wind can considerably damage a membrane, which has not been tensioned, and could even bring down an insufficiently stayed primary structure. Rainwater results in increased loading and can amount to several tonnes over a large area.

If the membrane surface and the primary construction stabilise each other in the structure, then the tensioning sequence for the introduction of the pretension must be designed according to the forces and deformations in the whole structure. The deformations arising in the structure from the staged tensioning are calculated as an iterative process in the erection planning.

The extent of the pretension to be applied at each stage must be specified so that the stability conditions in the structure during tensioning are compensated. Depending on the type of structure and the erection procedure, it is also necessary to define zones for variations from the relevant erection state. Loading on the structure resulting from geometrical imperfections during erection must be taken into account in the structural design.

When erecting flat and three-dimensional wire rope structures, the pretension is applied in stages (top in Fig. 208). The tensioning sequence is mostly specified as a star shape (bottom in Fig. 208). In some cases, it may be necessary to specify a reduction of the tension applied in some areas, as the slack and the pretensioned areas change places.

The erection of stabilised membrane surfaces is normally done alternatively in time and location, so that sufficient stability is present in the overall system at every stage (Fig. 209). After the hanging and connecting of the elements, a preliminary tension is applied in order to stabilise the surface against wind. The application of the final pretension to achieve the intended geometry is done after the installation of all membrane panels.

In Figure 210, constructional measures for stabilisation are derived from the type of loading on the structure and compared.

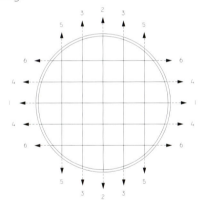
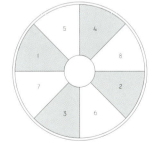

Fig. 208: Schematic diagram of a possible tensioning sequence

Fig. 209: Diagonally alternating installation of the Forum roof of the Sony Center in Berlin, Germany *(bottom)* and at the Millennium Dome in London, England *(top)*

1 Siokola, W. (2004)

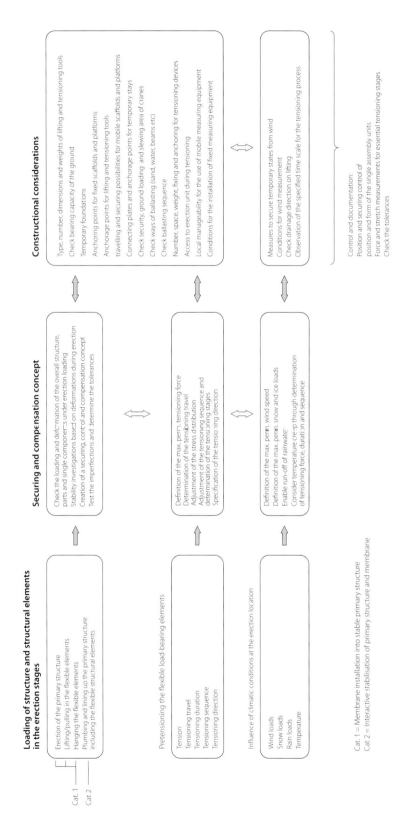

Fig. 210: Loading on the structure and stabilisation measures for form-active structures

3.4.1.3 Local conditions on the construction site

In order to ensure problem-free delivery and temporary storage and optimal conditions for unobstructed preassembly and erection, it is necessary to understand the conditions on the site exactly.

This can be gained by extensive local reconnaissance, documentation and surveying, and the following points should be investigated.

For the delivery of the construction materials, the access roads must be organised so that the traffic to the construction site is not obstructed. The unloading station should be kept free of stored material and site facilities. The areas for the temporary storage of the membrane should be paved, level and without steps, ditches or other obstructions. Storage areas for rope drums and membrane packets must be accessible for site vehicles (heavy trucks, mobile cranes, mobile platforms, fork lifts etc).

At the preparation area, the membrane panels are prepared for lifting or erecting and for pretensioning. The delivered membrane and edge rope and load-bearing elements of the primary structure are, depending on the erection procedure, laid out, connected to each other and prepared for lifting. This needs the area to be sufficiently large and flat. It should also be cleaned and laid out with a protecting foil, in order to effectively protect the sensitive fabric and wire rope surfaces from damage. It is also advisable to close off the preassembly area.

If membrane panels are to be installed into an already existing primary structure, then access must be ensured to the relevant locations for the erection work. There must be sufficient working space, especially at the edge areas where the pretension is applied and at the places where the membrane is to be lifted and unrolled or unfolded. There should also be space for the necessary platforms and press tables at the appropriate locations. For the installation of the membrane, it is also often necessary to lay hoses and electrical cables and to provide pumps. Any blower equipment and control desks should also be well accessible. When working at night, there should be adequate artificial light without shadows.

If scaffolds are to be used, it should be checked whether they are available on site and can be used. The aim in membrane erection is to make do with as little scaffolding as possible.

For longer-term erection works, tool and equipment stores and workshops for repairs are normally required. Meeting and social rooms and appropriate accommodation for the the erectors will also be needed.

The economic conditions at the construction site can also influence the choice of erection procedure. In densely populated regions of the third and fourth world, many workers are available with little expert knowledge. It can be a problem here to order the necessary lifting devices like cranes. Other construction infrastructure, tools and materials often only have a limited availability in such countries. In the highly industrialised countries, in contrast, it will be necessary to manage with a smaller number of highly qualified workers. It is therefore well possible that the implementation of the same construction task would be performed in very different ways at different locations. Transport, construction site set-up, completion deadline and finally also the construction costs can be considerably influenced by such limitations and obstructions.

Fig. 211: Conditions on the construction site and resulting erection conditions

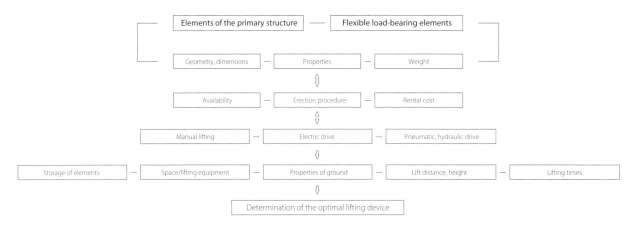

Fig. 212: Criteria for the selection of a suitable lifting device

3.4.1.4 Type of construction equipment

Lifting and tensioning tools and the load attachment equipment belonging to them are the main type of construction equipment used for the erection of membrane structures. The type, weight, number, loading and duration of use of the devices and equipment used determine to a considerable extent the effectiveness of the erection progress and thus also the cost of erection. The application of heavy lifting equipment in particular means high rental costs and requires the exact planning of the duration of use. The number and duration of lifting times and the resulting relocations, which become necessary, have to be kept within economic limits.[1]

The criteria for the selection of suitable lifting equipment can be summarised as shown in Figure 212.

The most commonly used heavy lifting devices in membrane construction are truck-mounted cranes with telescopic or lattice booms. In addition to the required load capacity and terrain capability, the reach and lifting height also need to be considered. The capability of extending the boom with additional sections and extending and withdrawing the telescopic boom under load also need to be investigated.[2]

Erection units and construction elements with smaller dimensions and lower weights can be moved, slewed, aligned or lifted with small lifting devices. The most commonly used tools, which are loaded in compression, are hydraulic lifting jacks, hoists and spindles; and those, which are loaded in tension, are block and tackles, chain hoists, tirfors and rope winches.

These small lifting devices are normally only suited for use in restricted areas due to their limited lifting height. The economy of their use depends on the ratio of force applied and dead weight to the load-lifting capacity.

Fig. 213: Use of cranes in membrane construction; *left:* Truck crane with telescopic boom, lattice-boom crane; *right:* Tower crane

---

1 *Wegener, E. (2003)* 2 *Bauer, H. (1991)*

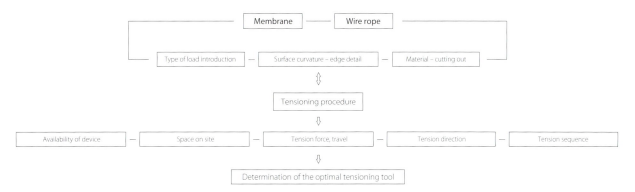

Fig. 214: Criteria for the selection of the correct tensioning tools

The tools and equipment used for tensioning membrane structures also influence the economics of erection. The relevant factors here are the type of equipment, weight, number, loading and duration of use. In discussing this equipment, it is sensible to distinguish between tensioning devices and equipment to erect wire ropes and devices, tools and temporary apparatus for the pretensioning of membrane panels.

Hydraulic presses and strand tensioners are the equipment mostly used to tension wire ropes with high forces. Tensioning of membranes is, depending on the tensioning process, usually done with hydraulic presses, tirfors, chain hoists or ratchet lever hoists, with the appropriate ancillary equipment.

The criteria for the selection of suitable tools for tensioning are summarised in Figure 214.

Tensioning equipment for wire ropes can have considerable weight and can then only be moved with a crane. The time taken to set up and relocate such an apparatus can be several days. This results in the following rule of thumb for tensioning wire ropes: the lower the number of tensioning set-ups required and the lighter and more manageable the equipment, the more economical is the erection.[1]

When considering tools for gripping the membrane edge for tensioning, it is not the weight and the related relocation and installation work that are decisive, but the number and type of the tensioning aids, which have to be made. The introduction of force using lightweight tools achieves more flexibility during erection. The elements gripping the edge of the membrane should be easy to install. A low number of components reduces the costs on the site and makes membrane construction more economical.

Hundreds of tensioning aids are often required for a single project. If the force is applied to the edge by successive tensioning and no appropriate constructional measures have been provided to assist the work, then these all have to be specially produced.

This makes it clear that the method of pulling the edge can be an economic question. The design of the edge detail and the tensioning concept should include discussions with the construction firm.

Fig. 215: *left:* Apparatus for tensioning stay rope; *middle:* Apparatus for tensioning carrier rope; *right:* Apparatus for tensioning stay rope

1 Inauen, B. (2003)

3.4.1.5 Methods of production, delivery and transport

One important advantage of building with lightweight structures in the high proportion of works prefabrication. In order to best exploit this advantage and the resulting short construction times, the largest possible membrane panels and rope lengths should be used.

Depending on the material, production method and details of delivery, fabric membranes can be delivered today in sizes of many thousand square meters. The size of each assembly unit and the material used essentially determine the manufacture and the delivery times for the materials. To obtain large areas exceeding the capacity of a single fabricator, it would also be conceivable to spread the production over several companies.

If fabric membranes are damaged during erection so badly that they cannot be rectified on site, then they will have to be newly fabricated again. Depending on the availability of the raw material and the manufacturing conditions, the delivery of new fabric pieces can take many weeks.

The production of several hundred metres of wire rope is quite usual for wire rope companies today. The limits of wire rope production are brought about internally by the capacity of the stranding machines and externally by the practicalities of loading and transport.[1]

If wire ropes are damaged, then the delivery of new ones can take many weeks. Rope manufacturers can also only deliver on time if sufficient wire is available.

The delivery of material from the fabricator to the construction site is the most important in the chain of erection processes. It is important that the material in this delivery is properly packed and protected from external influences. In addition to the packing, the method of erection is also significant. Rolled fabrics and foils are lifted and laid out in a different way to folded membrane panels. The elements of the primary construction, like masts, edge elements or rope bobbins, should be delivered as near as possible to the crane hook.

To determine the optimal transport quantities and means of transport, thee is a difference between transport to the site and transport on the site. The important factors for planning the transport are the

- location of the construction site,
- type and condition of the transport routes and distances to the site and on the site,
- dimensions, geometry, quantity and weight of the elements,
- availability of lifting devices,
- possibility of transport stabilisation,
- unloading conditions,
- legal regulations,
- transport costs.

Fig. 216: *left:* Mast transport; *middle:* Unloading a folded package; *right:* Unloading and lifting rolls

---

1 *Verwaayen, J. (2002)*

3.4.2 Remarks about the erection of the primary structure

Membrane structures have to be tensioned between fixed points, corners, edges or rings, where the loading is transferred into the compression members of the primary structure, which are mostly vertical, diagonal or arch-shaped. Because it is necessary to make mechanically tensioned membrane surfaces load-bearing through appropriate shaping, the structural form of the primary structure and its erection are also of special importance in the design, manufacture and erection. The most important primary structural elements for membrane structures are masts, edge beams and wire rope constructions. Some important aspects to be considered in the design, production and erection of these are presented here.

The stiff, linear edge beams and their supporting elements are mostly prefabricated and can have impressive size and weight, with the result that they have to be assembled on site. The dimensions of the prefabricated erection units derive from the transport capability and the space available on the preassembly area.

Procedures for the erection of stiff and flexible structural elements of the primary construction are illustrated in the following section with example projects.

3.4.2.1 Erecting masts

Load-bearing masts are the major element of primary constructions for high point structures or awnings. There are various methods of erecting masts vertically depending on the geometry, weight, availability of lifting devices and conditions on the construction site. The most common methods are summarised in Figure 217.

In a few cases, masts are also swung into position using rails or flown in and set up by helicopter.

Preparatory works

Before setting the mast in its intended position, various preparatory works have to be undertaken. Securing and stay ropes, lifting points and winches or pulley mechanisms can be prepared on the horizontal mast. If the flexible roofing surface is to be lifted at the same time as the mast, the membrane or rope nets must be laid out and mounted on the points provided (Fig. 218).

When segmental erection is used, the surfaces to be welded are usually conserved before welding. Markings enable precise levelling.

Lifting with one or more cranes

The lifting of columns in one piece to the final position is described as crane erection and is done using spreader beams or slings. When using more than one crane, these are usually the same type and have the same boom length and reach. To avoid risk of overloading and unintended lifting out of vertical, the lifting points need to be determined to distribute the load equally. The lifting process should take place as far as possible continually without impacts or swinging. Undesired load distributions can be avoided by planned and agreed working movements.[1]

The setting up of the 90 m long masts for the construction of the Millennium Dome in London (England) was an example for erection with more than one crane. 60 prefabricated mast sections were welded up with 480 joints.

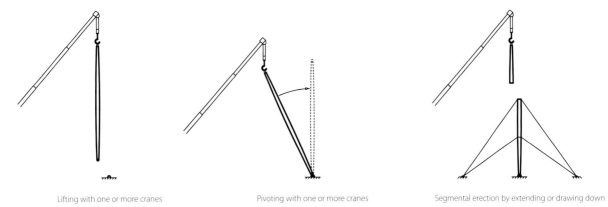

Lifting with one or more cranes   Pivoting with one or more cranes   Segmental erection by extending or drawing down

Fig. 217: Erection procedures for lifting columns into a vertical position

---

[1] Wegener, E. (2003)

Fig. 218: Preparatory work *left:* for pivoting; *middle:* for lifting; *right:* Pylon head with preassembled membrane

Fig. 219: Lifting the masts at the Millennium Dome in London, England

The 12 masts assembled in this manner were then driven by a 200 t crane from the preassembly area to the final location. Then the masts, each weighing 96 t, were lifted by a 1000 t crane with backmast (Fig. 219).[1]

Erection by pivoting using one or more cranes

The erection of masts by pivoting works on the principle of turning about an axis using a built-in hinge connected firmly to the foundation.

Fig. 220: Pivoting a mast with preassembled rope net

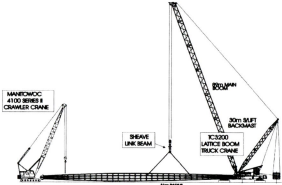
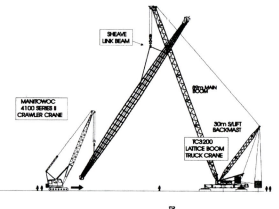
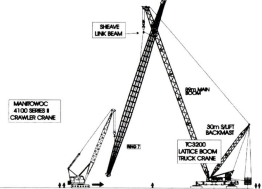

Fig. 221: Mast erection at the Millennium Dome

1 Miller, P. W. (2000)

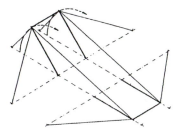 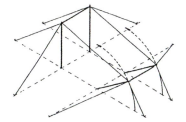 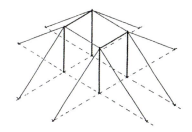

Fig. 222: Scheme of the mast erection of the 4-mast Chapiteau

Pivoting the mast vertical is mostly done with a crane. With smaller dimensions, lifting can also be done using A-trestles and a wire rope to pull vertical, or shearlegs with fixed rope guide. If rope winches are used to pull the mast up, they must be appropriately anchored. If the mast requires stiffening during erection, like a framed construction, then it is a good idea to combine stiffening and pivot hinge together.[1]

After being brought vertical, the column is aligned, temporarily supported on the foundation and lifted out of the hinge joint by mechanical or hydraulic lifting gear.[2]

One special form of pivoting vertical is used for the erection of circus tent masts. To erect the 4-mast Chapiteau, four steel tube lattice masts are usually laid opposite each other and pulled up with winches one after the other (Fig. 222).

In this very efficient process, first the anchor plates for the masts and stay ropes are set, then the two hinged masts are coupled at the mastheads with steel ropes. Electric winches installed at the foot points (d in Fig. 223) are used to vary the rope length between the opposing mastheads over deflecting rollers (e in Fig. 223).

After pivoting the first pair of masts into the intended position, the opposite masts can be hinged up by using the two synchronously connected winches (a, b, c in Fig. 223). The masts are supported by temporary stays against tipping sideways. The vertical position can be finely adjusted with tirfors between the stay ropes (f in Fig. 223). The setting up of the masts including all preparatory work can normally be done in less than 2 hours.

Fig. 223: Erection sequence and rope connections

1   Albrecht, R. (1973)
2   Seliger, P. M. (1989)

The speciality of this type of erection is not just the erection procedure, but also that the masts are reused many times. On account of the need in the circus world to roof over large areas in as short a time as possible, the masts are not only used as a load-bearing element in the completed state, but also as lifting device for dome construction, fabric membrane and electrical equipment during the erection.

Segmental erection through underpinning

The segmental erection of masts is an expensive process in terms of construction capacity, site infrastructure and erection costs. It is mostly only used for elements of large diameter, when lifting by crane is ruled out by the weight of the mast, or when space on the construction site is limited. It is necessary to build a special temporary construction for the assembly and welding of the mast sections and tensioning apparatus for pulling together and stiffening the sections. Another disadvantage is the high cost of scaffolding to create working platforms for working at great heights.[1]

The vertical erection of the sections can be done by extending upwards or by adding sections underneath. In the latter case, the top section of the mast is stood up in a temporary tower. After the top section has been jacked up, the next section is slid in underneath and welded. When erecting by extending upward, work starts with the base section. The following sections are lifted and set on top. In both cases, precise prefabrication of the sections is necessary. The length and weight of the individual sections have to be coordinated with the transport capacity (mostly maximum 12 m).

For the erection of the high point structure for a swimming pool roof in Kuala Lumpur, Malaysia, the approx. 100 m long main mast with diameter ranging from 900–3,100 mm was constructed by adding the sections underneath. The approx. 12 m long mast sections were welded in the vertical position in a temporary 30 m high assembly tower and lifted with four strand presses. Hydraulically controlled stabilising ropes held the mast vertical during the lifting process. After completion, the 300 t mast was set on a bearing ball with its cup and tilted to the correct position using the stay ropes (Fig. 224 and Fig. 225).[2]

To roof over the elliptically shaped Forum in the Sony Center in Berlin, a spoked wheel construction was chosen with spokes spread to an internal king truss supported by the cables.

The 4.000 m² roof has a span of 102 m over the major axis and 77 m over the minor axis. The roof surface is formed

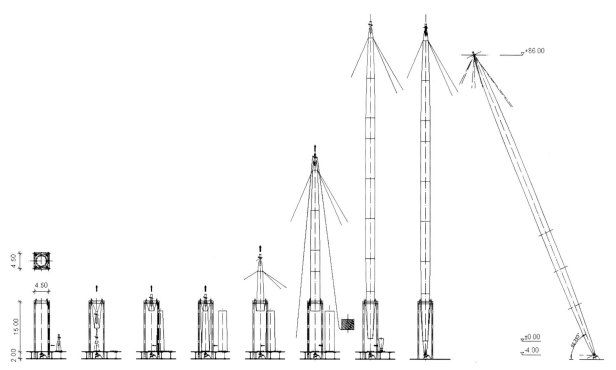

Fig. 224: Scheme of the mast erection for the swimming pool membrane roof in Kuala Lumpur, Malaysia

1 Wegener, E. (2003) *2 Schlaich, J.; Bergermann, R.; Göppert, K. (1999)*

Fig. 225: Segmental erection from underneath for the swimming pool membrane roof in Kuala Lumpur, Malaysia

by Glass/PTFE fabric and glass panels spanned between the ridge and valley ropes (right in Fig. 227).

For the construction of the king truss, a combined process of adding sections on top and underneath was used. The 45 m high, 8° tilted column with a weight of 100 t consists of 5 sections (right in Fig. 226), which were assembled in a 45 m high temporary scaffolding structure.

First, the completely preassembled central section was lifted into the scaffolding and provisionally anchored. Then the upper ring of the king truss was set, finely adjusted and welded. After the upper two sections of the king truss were complete, they were lifted 15 m and anchored to the scaffolding. Then the lower sections could be slid in, positioned and welded to the rest. Then the cable anchorage nodes for the foot point were lifted in and mounted. The lifting of the king truss was done using strand lifters, guide rails and armoured rollers mounted on the scaffolding. After the lifting process was complete, the king truss was swung into its final position with stay strands and the ridge and valley cables were mounted. The pretensioning of the system was done in partial stages over many days by telescoping the king truss at the foot point.[1]

  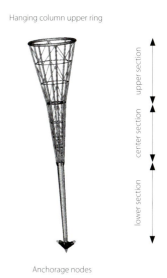

Fig. 226: *left:* Foot point of the king truss with cable anchorage nodes; *middle:* Head part of the king truss; *right:* King truss sections

1 Lindner, J.; Schulte, M.; Sischka, J.; Breitschaft, G.; Clarke, R.; Handel, E.; Zenkner, G. (1999)

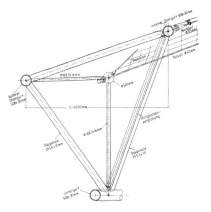  

Fig. 227: *left, middle:* Three-chord ring truss; *right:* Completed roof

3.4.2.2 The erection of edge beams

An example for the erection of edge beams for membrane structures is the assembly of the compression ring for the Forum roof described above. The structurally independent, pretensioned structure has an external compression ring constructed as a three-chord space truss ring beam (Fig. 227), which is supported at 7 points on the neighbouring buildings. It consists of 146 straight tubes for the chords and 361 tubes of various diameters for the struts. The tubes were delivered to the construction site singly and welded together there. Only the nodes with welded-on eye lugs and the ring beam bearing were made in the workshop. Because of its dimensions, it was not possible to manufacture the 520 t ring beams in sections in the workshop. Altogether 11 sections of 50 t each were assembled to form units, welded and tested at the preassembly area.[1]

An example of the assembly of a primary structure for membrane surfaces arranged over each other is shown by the segmental erection of the steel frame for the new 108 m high control tower at Vienna airport in Austria, where an external membrane envelope was built over a height of 45 m.

The Glass/PTFE fabric membrane panels are tensioned between 12 horizontal steel rings, which are connected by 8 spokes each to a central concrete shaft (right in Fig. 228). The total weight of the steelwork is about 200 t. The membrane span between the adjacent ring sections is about 4 m. The structure with a total surface of about 3.300 m² is the largest membrane structure in Austria.[2]

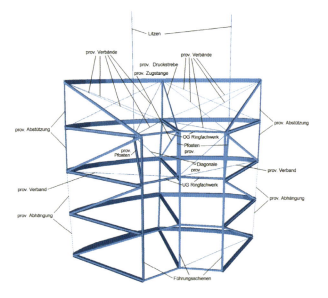 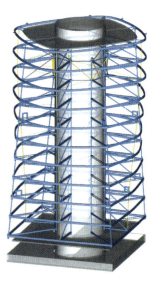

Fig. 228: *left:* View of a quarter of the steelwork with temporary construction; *right:* Steel construction around the reinforced concrete core

---

1 *Sischka, J.; Stadler, F. (2003)*

2 *Zechner, M. (2005)*

Fig. 229: *left, middle:* Erection of the first two steel rings; *right:* Completed membrane surface

The erection of the envelope was done by lifting. The uppermost three ring-shaped steel hollow profiles were assembled on the working platform at 28 m height complete with their bracing and hangers, and the upper three membrane fields were installed. Then the stiffened steel rings, supported at the eighth points, were lifted together with the two mounted membrane fields by 2 ring heights by four 700 kN strand lifters installed at a height of 71 m.[1]

The upper rings were stiffened for the lifting procedure for load distribution with a temporary structure, which was removed after completion. All the lower rings were hung from this "head area" with vertical tension members (left in Fig. 228). Brackets were welded onto the reinforced concrete core for the lifting procedure and guide rails mounted.

After the first steel rings had been lifted, the assembly of the next ring and the membrane belonging to it was possible. After coupling this to the completed segment, the construction was lifted by one ring height. This staged lifting process was continued until the entire membrane construction was in its intended location.

An example for a membrane structure with hybrid primary structure is the Palais Rothschild in the Vienna inner city. This has a combination of arch and wire rope construction as primary structure for a 18 x 15 m atrium roof with air pillow covering.

6 arches span over the atrium in the long direction and are supported transversely at the sides by horizontal framing. The arches, stainless steel profiles fully welded with X-seams at the works, have a rise height of approx. 170 cm and are stabilised from the side by diagonal compression struts, which stand on a horizontally spanned rope net. The rope net is tensioned against the horizontal framing at the sides and against steel profiles mounted on the reinforced concrete slab at the long sides.

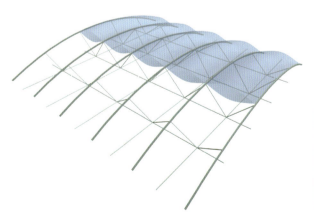

Fig. 230: Atrium roofing at the Palais Rothschild in Vienna, Austria; *left:* Scheme of structural system; *right:* Interior view (Photo © Werner Kaligofsky)

1 *Lorenz, T.; Mandl, P.; Siokola, W.; Zechner, M. (2004)*

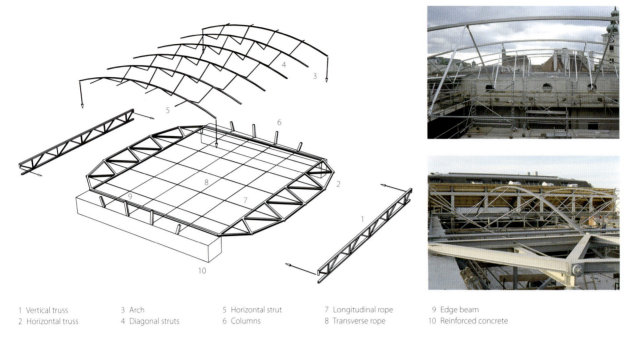

1 Vertical truss	3 Arch	5 Horizontal strut	7 Longitudinal rope	9 Edge beam
2 Horizontal truss	4 Diagonal struts	6 Columns	8 Transverse rope	10 Reinforced concrete

Fig. 231: Primary structure

Horizontal compression struts at the long sides stiffen the structure in the horizontal direction. The roof is covered with 5 inflated ETFE pillows spanned between the arches.

After the erection of the vertical and horizontal framing, the arches were lifted by a truck crane, bolted to the horizontal framing and stabilised with belts. After the installation and tensioning of the longitudinal wire ropes, the transverse ropes could be connected and the diagonal struts mounted on the ropes. The adjustment of the diagonal struts and the fine tensioning of the transverse ropes brought the structure into the final intended position.

3.4.2.3 Erection of wire rope construction

Pretensioned cable structures are often used as the primary structure for textile surfaces where the span is large and lightweight elements are desired. Pretensioned cable systems are often used above all for stadium roofs, which are covered with membranes.

Pretensioned cable structures normally consist of many single wire ropes, which are coupled with free-hanging node points or with elements, which are stiff in compression or bending. The erection of these structures is basically similar to that of single ropes.

Fig. 232: *left:* Hinged support on arch; *middle:* Connection of the diagonal struts to the longitudinal rope; *right:* Rope connection to the horizontal framing

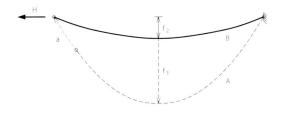
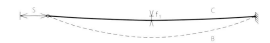

A Lifting phase	H Rope force
C Final position	f1 Sag in final state
S Tensioning travel	f2 Sag at start of tensioning
B Tensioning phase	f3 Sag at end phase of lifting
	a Rope shortening in lifting phase

Fig. 233: Erection of a single wire rope

The rope is first hung to a fixed point with large sag and then led to the tensioning point with a light force (A in Fig. 233). This lifting process can, for example, be done with a tirfor or a truck crane. The slack hanging rope can then be continuously tensioned by the appropriate tensioning tool until the intended tensioning travel (S in Fig. 233) has been reached and the intended rope geometry and force have been achieved. In principle, it should be noted that ropes are mostly shortened at their end points. If this is not possible, then tensioning is done in the free length.

This erection process can essentially be divided into the lifting and the tensioning of the rope. While being lifted, the rope experiences mostly geometrical shortening, and while being tensioned mostly elastic shortening. Regarding the expense of time and equipment, it should be noted that the tensioning travel depends on the initial force on hanging the rope. When the initial rope length is shorter, the phases "lifting" and "tensioning" cannot be clearly separated on account of the force, which has to be applied; they overlap seamlessly.[1]

On lifting, the installation length of the rope is shortened. The geometrical change of shape, the force to be applied and the elastic deformation of the rope are small. On tensioning the rope, it is shortened further. In the initial phase, the geometrical change of length and the elastic deformation of the rope are approximately of the same order as those on lifting. In the final phase of the tensioning process, where the rope nears its intended length, the tensioning force multiplies many times. Predominantly elastic extension of the rope is caused. Tensioning devices must often be first re-equipped or connected during this phase.[2]

The erection principle for the installation of a single rope can basically also be used for the installation of planar and spatial rope structures.

To achieve the intended shape of planar pretensioned structures, the type of force introduction has to be considered. For tensioning rope trusses, this can be shown by reference to two methods. If the bearings are easily accessible, then the structural system can be efficiently pretensioned by shortening the spanning ropes. If the ends of the spanning ropes are not accessible or poorly accessible, then the hanger ropes can be shortened. This does require less force to be applied, but the method demands more work and more use of equipment.

Shortening of the hanger ropes

Shortening of the spanning ropes

Fig. 234: Tensioning a planar rope truss

---

1 Kleinhanß, K. (1981)

2 Kleinhanß, K. (1981)

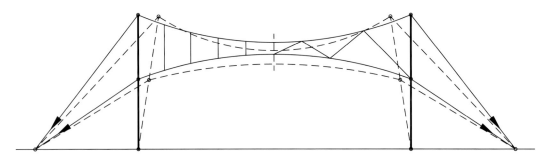

Fig. 235: Pivoting the masts vertical by shortening the stay ropes

Planar or spatial rope structures with load-bearing members stiff in compression and bending can also be tensioned by sliding the rope abutment if the connection nodes are poorly accessible. For rope systems with back-anchored masts, this can be achieved by pivoting the masts on hinged bearings. To erect the masts and lift the carrier and stay cables, temporary cables are pulled in and tensioned to sufficiently stiffen the structural system and stabilise it sideways.

Wide-span spatial rope systems are a very efficient primary structural system for membrane roofs, seen from structural and economic points of view. A characteristic example for the design and erection of such a system is a rope net system patented by David Geiger.

Ridge ropes arranged in rotational symmetry in this dome-shaped system span from an external compression ring to an internal tension ring. The ridge ropes are supported underneath spatially by compression struts. The ropes supporting these compression elements form, on the one hand, horizontal, polygonal compression rings and, on the other hand, diagonal ropes in the vertical plane of the ridge ropes (Fig. 236).[1]

The assembly of the rope dome is illustrated in Figure 236. The radially arranged ropes laid out on the ground are connected to the head of the compression struts. The inner steel tension ring is set up and bolted to the radial ropes.

After the foot points of the struts have been connected with the strands of the polygonal tension ring and the diagonals, the feet of the struts are connected to the radial ropes. Then first the outer and then the inner diagonals can be successively tensioned until the ring reaches its intended position.

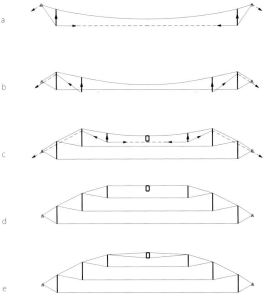

Fig. 236: Wire rope structure, Geiger system

a – e = Erection principle

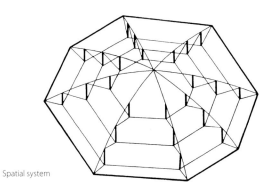

Spatial system

[1] Stavridis, L. (1992)

One type of structure, which produces no horizontal forces externally under vertical loading internally, is the constructional principle of the spoked wheel. The structural system, described as a bicycle wheel roof, is mostly used for the membrane roofing of sports stands.

The chief characteristic of the constructional system of a spoked wheel is that an internal tension ring is braced to an external compression ring by spokes splayed either outwards or inwards. The structural behaviour is determined by loads acting perpendicular to the plane of the wheel being resisted by diverted forces from the rings so that the system transfers no horizontal forces to the outside. It should be pointed out that asymmetrical loadings are problematic for spoked wheel constructions, because they can cause great deformations and the constructional principle can then be poorly exploited.[1] Any variation form the circular form must therefore be very carefully considered, just as an opening of the inner ring or geometrical differences between inner and outer rings.

The erection of spoked constructions is essentially determined by the nature and the location of the forces applied. For the version with internal node point and spokes splayed outwards (A, a1 in Fig. 237) or inwards (a2 in Fig. 237), the planes of the ropes can be pushed against each other by centrally placed press equipment. For constructions with an inner ring, the loads are normally transferred through tension connections mounted on the compression ring, i.e. peripherally (b1 in Fig. 237), in which case it is possible to tension synchronously and under central control with presses arranged at the node points. One exception to this is a tensioning method where the pretension is introduced by the lowering of the inner ring (b2 in Fig. 237).

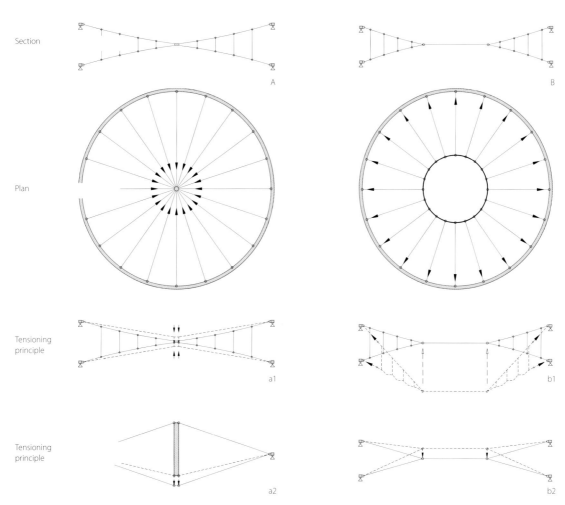

Fig. 237: Method and location of force introduction into bicycle wheel structures

---

1 Reitgruber, S. (2003)

Fig. 238: Primary structure of the Gottlieb Daimler Stadion in Stuttgart, Germany

Druckring-Oben-Aussen
Stahlhohlkasten 1200*1200 mm
Länge 21 m

Druckring-Oben-Innen
Stahlhohlkasten 600*900 mm
Länge 21 m

Druckring-Pfosten FL 60*900
Druckring-Diagonalen 2 | 50 mm

Druckring-Unten
Stahlhohlkasten 900*900
Länge 20 m

Stützen
Stahlhohlkasten 550*550-1100
Länge von 28 bis 49 m

Seilbinder
Tragseile von VVS | 71 - VVS | 99 mm
Hänger Litze | 22 mm
Spannseile von VVS | 61 - VVS | 74 mm

Ringseil
2*4 VVS | 79 mm

Girlandenseile
Litze | 44 mm

Fig. 239: *left:* Primary structure with rope trusses; *middle:* Lifting the rope structure; *right:* Erection of the compression ring

No example of simultaneous application of the tensioning forces by increasing the circumference of the edge beam, as is the case with saddle rope net structures, is known to the author.

When constructing ring cable roofs for stadium roofing for covering with membranes, the erection of the flexible structural parts of the primary construction is one of the most important erection procedures. The ring rope bundle and the stay, carrier and hanger ropes are preassembled on the ground, pulled up together and tensioned against the compression ring. Then follows the erection of the membrane panels, which form the secondary system together with their edge elements.

Important experience was gained with this erection procedure by the design engineers and construction firms involved during the construction in 1992 of the Gottlieb Daimler Stadion in Stuttgart, which was then the largest membrane roof in the world. The two steel elliptical perimeter compression rings are supported on 40 steel columns at a spacing of 20 m and form the primary structure together with 40 spoke-shaped cable trusses running inwards (Fig. 238 and left in Fig. 239). The secondary system consists of arches mounted on the carrier cables with membrane panels spanned between them. The primary structure is stable with the cables crossing at the quarter points of the perimeter between the columns and the provision of large foundations; the framing by the roof envelope is not required for structural stability.[1]

After the preassembly and lifting of the assembly units of the compression rings (right in Fig. 239), the lifting of the cable construction began. The 40 cable trusses laid out on the stand were lifted up together with the connected ring rope bundle laid out on the 400 metre track. This was done by 40 centrally controlled presses mounted on the compression ring, with a press travel of 15 m (middle in Fig. 239). Then the carrier cable could be hung.[2]

To tension the cable structure, the carrier cable was tensioned with a travel of 1.4 m. The press force in the curve truss increased at this stage from 450 kN to 2,300 kN (right in Fig. 240). The decisive point for the success of the lifting and tensioning of the cable structure was the simultaneous tensioning of all trusses. The lifting and tensioning procedure was accompanied by checks of force and geometry and could be completed within 3 weeks.

Fig. 240: *left:* Schematic illustration of the lifting and tensioning procedure, *right:* Force curve in the carrier cable during the lifting and tensioning procedure

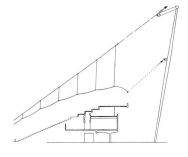
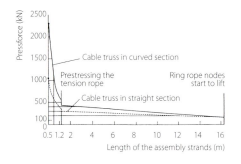

---

1 Bergermann, R.; Göppert, K.; Schlaich, J. (1995)

2 Bergermann, R.; Göppert, K.; Schlaich, J. (1995)

3.4.3 Erection procedures for membrane structures

The development of an erection principle and its implementation as an erection procedure requires a range of considerations regarding methods, processes and implementation of the necessary working steps.

In addition to the factors from design and detailing and the particular characteristics of material and production discussed in chapter 2, progress without disruptions on the construction site is also very important. This can be achieved by efficient organisation of the erection.

An important part of this is the scheduling of the construction, that is the sequence of works. The factors influencing the coordination of the working activities for the implementation of membrane construction are shown diagrammatically in Figure 241:

The following sections discuss the principles of erection and possible site procedures for the erection of characteristic structural forms in mechanically tensioned membrane construction.

Criteria affecting material, structural system, implementation of erection and the equipment used are shown in tabular form. Each construction phase is illustrated with diagrams and the procedure commented. Finally, there are pictures of completed projects. The projects chosen are only examples; the list does not claim to be complete.

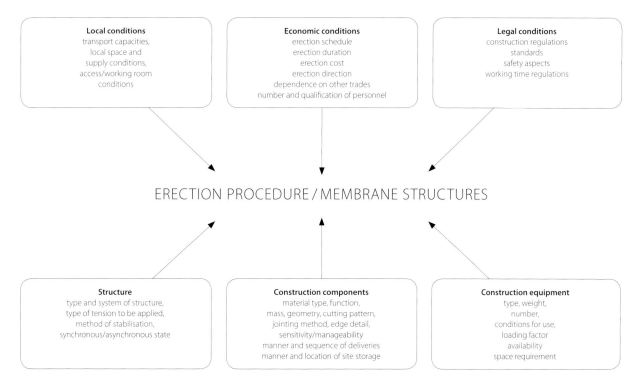

Fig. 241: Factors influencing the erection procedure

3.4.3.1 Erection of Awnings

Awnings	Material, structural system	Construction	Equipment
	Membrane material, cutting direction: PES/PVC fabric, parallel	Stabilisation of primary structure: crane, temp rope, belts	Type of scaffolding: truck-mounted working platform
	Support: at points, masts	Packaging of membrane: package/folded	Lifting device: crane
	Edging: flexible/rope edge, webbing edge	Load introduction: at points, corner fittings	Tensioning tools: tirfor, chain hoist, lever hoist

The erection of a four-point awning starts with setting out the foundation centres and bolting the masts. Then the stay ropes are attached to the mast heads.

After the membrane has been unfolded and laid out, the edge ropes are slid in and the corner fittings assembled. The fixing of the laid-out membrane surface to the primary construction is done by connecting the corner fittings to the mast head.

The structure is erected by successively pivoting up the masts. The masts have to be stabilised during this procedure.

Then the stay ropes are connected to the brackets on the foundations.

After the stay ropes have been shortened, the force is introduced into the membrane surface. The surface can be finely tensioned with the connection of the edge ropes to the corner fittings. When using Glass/PTFE fabrics, special care needs to be taken at this stage that the brittle fabric is not damaged.

If a number of high points are arranged at a sufficient spacing, then the opposing masts can be swung up cross-wise starting from the middle. Care needs to be taken here that the cranes do not lift obliquely.

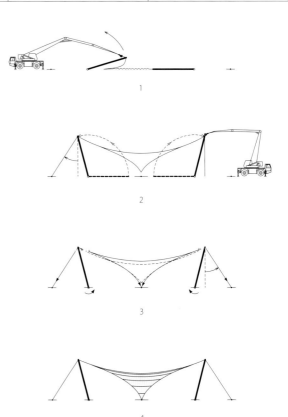

Fig. 242: Erection procedure for four-point awning

  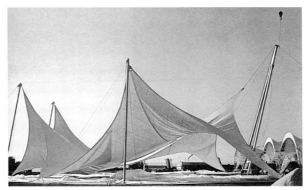

Fig. 243: Erection of a multi-point awning

3.4.3.2 Erection of high point surfaces

High point surface	Material, structural system	Construction	Equipment
	Membrane material, cutting direction: PES/PVC, Glass/PTFE fabric, radial	Stabilisation of primary structure: crane, temp. rope, belts	Type of scaffolding: truck-mounted working platform
	Support: from inside, at points, hanging masts	Packaging of membrane: package/folded, gathered	Lifting device: truck crane
	Edging: flexible/rope edge, webbing edge, clip edge	Load introduction: at points, corner fittings	Tensioning tools: hydraulic press, tirfor, chain hoist

If one or more points are arranged within the membrane surface where the surface is drawn upwards, this is called a high point surface. The high points are normally constructed as rings or boxes. They can be hung from outside or from inside by standing or hanging masts. Their edges can be detailed flexible or stiff in bending.

In the following section, various principles for the erection of high point surfaces are presented, covering those supported internally or hung externally, with a single high point or compartmented.

A very efficient method of erecting single or coupled high point surfaces with hanging struts supported underneath by ropes and with edge ropes, is the simultaneous lifting of primary structure and membrane.

Firstly, meticulous preassembly is required. The edge and hanging struts are laid out on the ground and the membrane is fixed to them in a way that keeps it free of damage during the lifting operation in a temporarily tensionless state. Careful attention needs to be paid to the design of the clamping of the high point and the assembly joints for the adjacent membrane panels.

After lifting the high point with a truck crane and hanging the supporting rope, the edge support struts can be turned to the outside and temporarily fixed with ropes of belts. The hanging struts can be elongated using presses and the supporting ropes are fixed by the post-tensioning.

The edge and stay ropes are then tensioned in stages until the calculated level of pretension is reached.

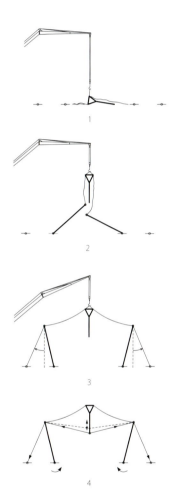

Fig. 244: Erection procedure – high point surface

Fig. 245: Erection of an internally supported high point surface

High point surface	Material, structural system	Construction	Equipment
	Membrane material, cutting direction: PES/PVC, Glass/PTFE fabric, parallel	Stabilisation of primary structure: crane, temp. rope, belts	Type of scaffolding: **system scaffolding, truck-mounted working platform**
	Support: from inside, at points, masts	Packaging of membrane: package/folded, gathered	Lifting device: truck crane
	Edging: flexible/rope edge, clip edge	Load introduction: at points, corner fittings	Tensioning tools: hydraulic press, tirfor, chain hoist

For the erection of internally supported, compartmental high point surfaces, it is sensible to first erect the edge columns in their intended positions, secure them with temporary ropes or belts and fix the corners of the membrane to the heads of the columns.

This does, however, require scaffolding on which to lay out the membrane.

The pulling in of the edge rope and the fixing of the corner fittings prepares the membrane surface for lifting. After fixing the membrane to the columns and making the sectionalising joints, the scaffolding can be dismantled.

The highpoint now hangs loose underneath, which is advantageous for drainage if it rains during erection.

To mount the hanging columns, they can be stood in position on the ground. After fixing the membrane, the columns can be successively lifted from the outside or from the inside and the membrane pushed out to form the high points. This does, however, require a sufficiently large spacing of the high points and a sufficiently large high point radius. Particular care should be taken at this stage with Glass/PTFE fabrics to avoid damage to the coating.

A slight lifting of the hanging columns enables easier hanging of the supporting ropes.

Pushing up the hanging columns in stages with hydraulic presses effects the force introduction into the membrane surface. The surface can then be fine tensioned panel by panel by shortening the stay ropes and tensioning the edge ropes.

1

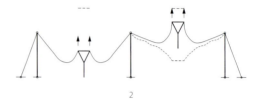

2

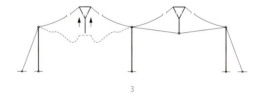

3

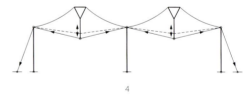

4

Fig. 246: Erection procedure – high point surface

Fig. 247: Erection of the vehicle park roof at the Munich Waste Disposal Office

An example for the erection of an internally supported high point surface with many high points is the construction of the roof over the stand at the Jaber Al-Ahmad Stadium in Kuwait.

The perimeter rope roof is, like several other stadium roofs with membrane covering, constructed according to the principle of a spoked wheel, but the design and construction as a single-layer rope net construction without rope trusses splayed inwards or outwards was an innovation for this roof. A strongly crimped perimeter beam acts as a compression ring, from which radial ropes span to the internal, geometrically affine rope tension ring. Together with the groups of ropes running perpendicular to the radial ropes, this forms a spatially curved system, on which the roof envelope of Glass/PTFE membrane sits. The membrane panels span between the radial ropes and are pushed up by the compression struts, which are supported underneath by ropes at 7–9 high points per field in order to achieve surfaces with curvature in opposing directions (Fig. 248). The edges of the fabric have clamping plates, which are fixed to the radial ropes with clips.

After the erection of the perimeter rope bundle and the rope net has been laid out and preassembled, the radial ropes were lifted with strand tensioners over many weeks and bolted to the compression ring (Fig. 249). The erection of the membrane roof was then done in panels next to each other (Fig. 250). After the hanging struts had been preassembled and secured, the fabricated membrane package was lifted with tower cranes. Working from a platform hanging from the crane, the packages were folded out, clamped to the high points and fixed to the radial ropes with temporary clips (Fig. 251).

The pretensioning of the hanging struts was done in many stages. First, the high points were brought into position with chain hoists. After the fabric had been fixed to the high points and the edges (Fig. 252), the hanging struts were lightly tensioned field by field, before being brought into their final position in a further perimeter sequence (Fig. 253).[1]

Fig. 248: View of the roof with high point surfaces

Fig. 249: Single layer ropenet construction

Fig. 250: Lifting the last membrane panel

Fig. 251: Pulling the membrane off the package from a temporary platform

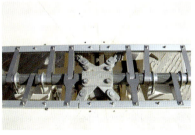

Fig. 252: Edge detail with clips and cast nodes for the ropes spanned underneath

Fig. 253: Application of the pretension

1 Lenk, S. (2006)

High point surface	Material, structural system	Construction	Equipment
	Membrane material, cutting direction: PES/PVC, Glass/PTFE fabric, parallel	Stabilisation of primary structure: crane, temp. rope, belts	Type of scaffolding: truck-mounted working platform
	Support: from inside, at points, masts	Packaging of membrane: package/folded, gathered	Lifting device: truck crane, winches, pulleys
	Edging: stiff/rope edge	Load introduction: at points, corner fittings	Tensioning tools: hydraulic press, tirfor, chain hoist

High point surfaces, which are supported on internal masts, can also be erected very efficiently.

This is done using a technique, which comes from circus building.

The masts can be stood up as described in section 3.4.2.1 by opposing pivoting. The arrangement of the winches at the foot points of the masts and sheaves at the mast heads are used to lift the membrane surface.

After the membrane has been laid out in a protected area, the edge ropes and the corner plates can be premounted and the stay poles connected.

After the preassembly, the pulling up of the membrane along the temporarily stayed masts can be begun.

During the lifting process, the stay poles can be pivoted one after the other with cranes and fixed with rigging stays.

The entire lifting process, depending on the number of edge columns, the material used and the stress distribution in the surface, is normally completed in a few days.

The planned pretension is reached with the tensioning of the stay ropes and the edge ropes. The tensioning procedure is constantly monitored with force and geometry checks.

1

2

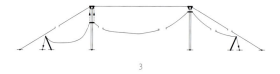
3

4

Fig. 254: Erection procedure for a high point surface

Fig. 255: Erection of a high point surface with internal masts

High point surface	Material, structural system	Construction	Equipment
	Membrane material, cutting direction: PES/PVC, Glass/PTFE fabric, parallel	Stabilisation of primary structure: crane, temp. rope, belts	Type of scaffolding: truck-mounted working platform
	Support: external, at points	Packaging of membrane: package/folded, gathered	Lifting device: truck crane, winches, pulleys
	Edging: flexible/rope edge	Load introduction: at points	Tensioning tools: hydraulic press, tirfor, chain hoist

If the room under a high point roof is required to be free of columns or masts, it can be hung from the outside through plates or rings.

If the external primary structure consists of stayed masts, then the membrane surface can be hung from a cable hanger system. Diagonal ropes with pulleys at the mast heads can be used as lifting devices for pulling up the roof envelope. When using this method of lifting, it is important to pay attention to the angle of the rope axes and the geometry of the high points.

The two masts can be stood up with truck cranes, in which case the carrier cable must be lifted together with the preassembled hanger ropes.

If the mast stays are fixed, the membrane surface can be laid out on the ground. After the high points have been preassembled and the assembly joints have been made, the corner plates can be mounted and the edge rope pulled in.

Using electric winches and sheaves mounted on the mast heads, the roof envelope can be pulled up.

After the edge columns have been turned out, the high points can be connected to the hanger ropes.

The pretension can be applied by shortening the stay ropes and pivoting the masts.

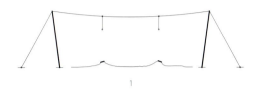
1

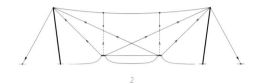
2

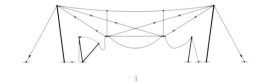
3

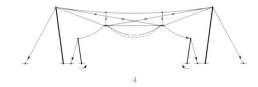
4

Fig. 256: Erection procedure for high point surface

Fig. 257: Erection of a hanging high point surface by pulling up the hanger rope system

An example for the erection of a high point construction supported by an external rope net is the construction of the Velodrome (Redesign) in Abuja in Nigeria in 2006, an approx. 13,000 m² compartmental high point surface composed of pentagonal and hexagonal panels.

8 Pylons with a height of 50 m support a rope net (Fig. 258), which carries the envelope of Glass/PTFE fabric. The honeycomb panels, produced with a parallel cutting layout, are joined together along their edges by edge ropes. The external corner points of the perimeter fields are supported by 16 inclined columns.

To prepare for erection, the individual panels are laid out on the ground (Fig. 259), and fitted with temporary tensioning fittings at the edge and reinforcement at the corner cut-outs. Once this preparation work is complete, the panels are fixed to specially prepared hanging scaffolds, with which they can be lifted at any angle to the cast iron rope nodes using rope winches and pulleys. The arrangement of the hanging scaffolds along the edges of the surface enables the membrane panels, which are at risk of kinking, to be lifted while almost fully extended (Fig. 260).

After the high points have been bolted and lightly pretensioned, the adjacent membrane panels are joined to each other using a specially constructed temporary construction. Then the edges of the panels are pulled together with lever hoists or tirfors and clamped with temporary plates to the edge rope (Fig. 261). After erection of the edge panels (Fig. 262), each individual high point of the construction was pretensioned with hydraulic presses exerting approx. 15 t (Fig. 263).

Fig. 258: Rope net construction

Fig. 259: Preassembly of membrane panel

Fig. 260: Lifting up a membrane panel

Fig. 261: Connecting a membrane panel

Fig. 262: Erection of edge panels

Fig. 263: Application of the pretension

3.4.3.3 Erection of arch surfaces

Membrane surfaces supported in an arch-shape are a very efficient structural system and are often used as a structural form in membrane construction.

When a number of arch fields are arranged together, the membrane surface can have, through the interaction with the arch determined from the ideal supporting form, a stabilising and a loading effect in the structural system at the same time and thus effectively secure against buckling.[1]

When erecting arch surfaces, the strip layout and the form of edging determine the installation direction. If the warp direction runs parallel to the arch and if the membrane surface is linearly clamped to the arch, then it can be tensioned perpendicular to the arch (1 in Fig. 264). This is done by clamping the membrane in one arch and successively pulling and clamping to the opposite arch, starting from the middle. The edge is finely tensioned by tensioning the edge rope. The typical arch edge detail for this strip layout is the clamping plate. The commonly used tensioning tools are tirfors or lashing strap systems with ratchet tensioner.

If the membrane surface is clamped to the arch at points, then it can be tensioned by moving the corner fittings and shortening the edge ropes (2 in Fig. 264). When using this method of edge detail, great care should be taken when tensioning no to overload the tip of the membrane. The stress distribution in the membrane is the most important factor in planning the tensioning sequence and the tensioning force to be applied. Arch membranes clamped at points can also be unrolled perpendicular to the arch direction.

If the warp lies perpendicular to the arch and the membrane is linearly fixed to the arches, then the membrane can be installed very efficiently if the edge detail is appropriate (3 in Fig. 264). If the structural requirements resulting from load transfer allow it, it is advisable to detail the stiff membrane edge with a keder rail. After clamping the membrane at both ends of the arch, the keder can be drawn through the keder rail running in the direction of the arch and tensioned. This method of erection is normally very fast and takes less time than the process described above. But it is not possible to use light mechanical tensioning devices to tension on account of the long travel and the high forces to be applied. Heavy electric rope winches are normally used.

Less often, the membrane surface unrolled perpendicular to the arch direction is also tensioned to the opposite arch (4 in Fig. 264). When tensioning in the stiffer warp direction and along seams, care should be taken with the level of the tension forces applied.

Other methods of applying load into arch surfaces, like for example the lifting or pivoting of the arches, are described in section 3.4.3.

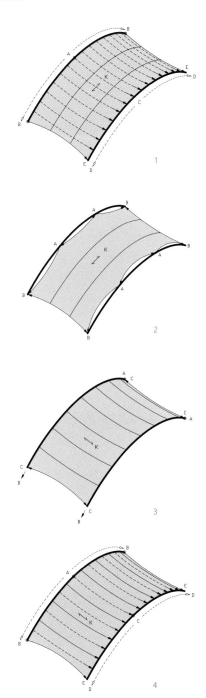

Fig. 264: Erection of arch surfaces

---

1 Alpermann, H.; Gengnagel, C. (2003)

High point surface	Material, structural system	Construction	Equipment
	Membrane material: PES/PVC, Glass/PTFE fabric	Method of stabilisation: temp. rope, belts	Type of scaffolding: hanging scaffold, abseil working
	Support: linear	Packaging of membrane: gathered longitudinally or rolled	Lifting device: crane with spreader beam
	Edging: clamping plates, rope in sleeve	Load introduction: linear, successive	Tensioning tools: hydraulic press, tirfor, belts

Arch membranes are frequently installed as ring rope roofs for stadiums, with the arch being supported with a hinge on the carrier cable and the warp direction of the fabric mostly running at right angles to the arch. When several arches are arranged next to each other, the membrane is mostly pulled in along the arch from rope truss to rope truss.

Therefore it is a good idea to fabricate a keder into the underside of the membrane and pull the fabric transverse to the field along a keder rail intended to provide resistance against wind uplift mounted over the length of the arch (4 in Fig. 265). When doing this, the friction resistance in the keder profile should be paid attention to.

The installation of the arch and the pulling in and tensioning of the membrane are mostly done field by panel from scaffolds hanging under the rope trusses or by abseiling. To stabilise the arches, temporary ropes tensioned diagonally in three dimensions can be installed during the erection (S in Fig. 265).

The rolled or gathered panels can be lifted with a rope spreader beam hanging from a crane and temporarily stored at the erection height (1 in Fig. 265). If this is not possible and the tensioning is to be done at right angles to the arch, the fabric is mostly preassembled along the arch (left in Fig. 266).

When pulling in the membrane in the arch direction, it is first anchored to the end points, in order to mount the edges to the truss. After pulling the membrane to the opposite side, it can be successively tensioned against the truss starting in the middle and fixed by mounting the edge elements (2, 3 in Fig. 265).

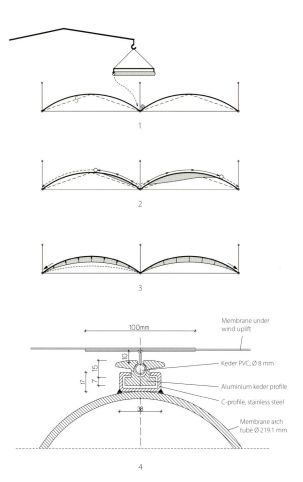

Fig. 265: Erection sequence – Arch membrane and detail of arch crown

Fig. 266: *left:* Membrane preassembled on the arch; *middle, right:* pulling in along the arch

An example for the erection of an arch structure with the surface unrolled transverse to the arch direction is the construction of the roofing over the platform at Fröttmaning station near Munich. The roof surface used was a Glass/PTFE fabric.

The load-bearing structure consisting of arched trusted spans over the platform and carries secondary arches on hinged supports, between which the membrane is tensioned. The warp direction runs parallel to the arch trusses.

The fabric, folded longitudinally, was lifted transverse to the arch trusses and unfolded on both sides over the secondary arches. (1 in Fig. 268). A net mounted under the arch truss served as a safety net.

After fixing the membrane to the corner points, the clamping plates were installed along the edges of the arch trusses (2, 3 in Fig. 268). Immediately after this, the edge ropes were bolted at the face side, which would have been very difficult at a later time.

The introduction of the pretension was done at right angles to the arch truss by moving the preassembled keder rail sections. This was done with the help of lashing straps, with the force being introduced successively starting from the middle of the arch truss (Fig. 267).

After the edge rope, which had already been pulled in, had been tensioned, the edge of the surface was finally tensioned through the webbing fabricated into the corners and fixed by bolting the ropes intended to secure against wind uplift to the crown of the arch (4 in Fig. 268).

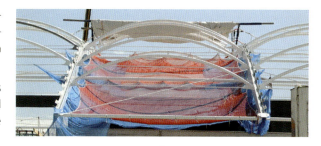

1

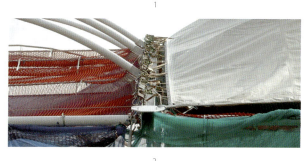

2

3

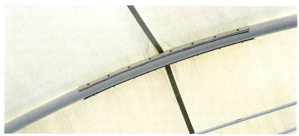

4

5

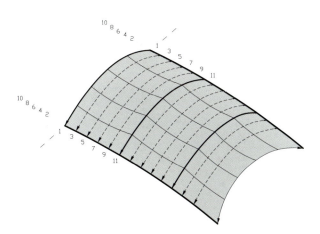

Fig. 267: Tensioning sequence and tensioning direction

Fig. 268: Erection of arch membrane at Fröttmaning station, Munich, Germany

Arch surface	Material, structural system	Construction	Equipment
	Membrane material: PES/PVC, Glass/PTFE fabric	Method of stabilisation: temp. ropes	Type of scaffolding: abseil working
	Support: linear	Packaging of membrane: gathered	Lifting device: crane with spreader beam
	Edging: clamping plates	Load introduction: linear, swinging the arches	Tensioning tools: tirfor, lashing straps

One way of roofing over large plan areas is to design an arch structure stayed sideways.

The staying can be, depending on the length of span and the rise of the arch, performed by rope bundles or membrane surfaces, with the result that the staying elements and the arches on hinged supports stabilise each other. If the saddle-shaped surfaces are spanned by ropes, they can be covered with membrane panels.

The membrane panels can be attached to the arches on the ground before erecting the primary structure. If this is not possible, then the panels must be lifted in (a, b in Fig. 269). To do this, it is advisable to provide in advance a belt net as surface to lay out the membrane.

For the erection of the membranes, the arches must be stabilised with temporary rigging stays. The arches are often pulled slightly together, in order to relax them again after erection (Fig. 269 c). The panels should be gathered for lifting, so that they can be laid out on the belt net without problems.

After the prestretching and clamping of the edge elements along the arches, the pretension can be applied by successively tensioning the compensated membrane edges.

Alternatively, the load can also be applied to the crown area by rotating the arches (Fig. 269 d).

After prestretching and clamping the side panels along the arches, their edges can be tensioned to the lower edge beams using tirfors (Fig. 269 e).

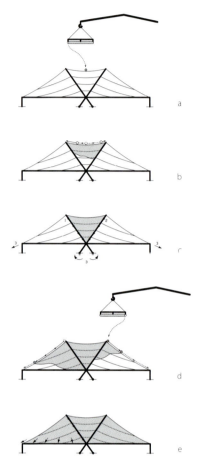

Fig. 269: Method of erection of arch surface

Fig. 270: Erection variant for arched construction

With a surface area of approx. 108,000 m², the roofs of the approx. 3,100 m long access concourses of the Suvarnabhumi International Airport in Bangkok are currently the largest membrane structure in the world (Fig. 271).

The concourses consist of 104 identical three-chord truss beams inclined towards each other in pairs, between which a three-layer membrane construction is tensioned with a span of approx. 27 m in the long direction of the building. The side panels located between them are glazed.[1] The outer of the three membrane layers consists of a Glass/PTFE fabric, the middle layer is a steel rope net with polycarbonate boards. The inner layer consists of a partially coated (partially open-pored) glass fibre fabric, which was coated on the inside with a vapour-deposited aluminium layer. This coating, described as low-E (low emissivity), functions as an infra-red mirror and hinders the exchange of heat radiation between the warm outer layer of membrane and the construction components of the inhabited zone inside the building.[2] The special feature of this three-layer construction, a new development which has been patented worldwide, is that in addition to fulfilling the energetic requirements for the space, the inner layer in combination with the polycarbonate boards of the middle layer achieves a noise reduction effect at the same time as being sufficiently translucent.

After the installation of the textile hanging scaffolds as access to the connection points along the trusses, a textile belt net with PVC membrane covering was spanned for the erection of the three-layer roof envelope. Firstly, the erection of the middle layer was begun with the steel rope net (Fig. 272). After the erection of the rope net, the polycarbonate boards were bolted to the cast iron nodes of the rope net and the joint seals were fixed (Fig. 273). Then followed the lifting of the approx. 1,000 m² membrane panels of the outer fabric membrane, whose cutting direction (warp) runs parallel to the axis of the building (Fig. 274). The Glass/PTFE fabric was first tensioned in warp direction with the bolting of the 4 membrane corner fittings. The tensioning of the outer layer in its final position was done in warp direction using ratchet tie-downs (Fig. 275). To fix the outer membrane, the membrane edges, detailed as clip edges (aluminium-clamping plates, stainless steel clips) were mounted on lengths of tubing, which are welded to the appropriate lower chord of the trussed arch. After welding the covering membrane (Fig. 276), the assembly of the outer layer was complete and the membrane panel watertight. As an edge detail for the inner layer of membrane, "claw profiles" were pulled onto their keder edge, which are fixed to the brackets on the steelwork with specially made shroud tensioners. The erection of the inner layer then completed the assembly of one of the 108 membrane panels.[3]

This erection procedure was repeated for each of the 108 membrane panels. 5 experienced membrane specialists managed 6 erection crews with 70 workers each working independently of each other, who, split up into working gangs, worked on six panels simultaneously in a rolling schedule.[4]

Fig. 271: Aerial view of the airport in Bangkok, Thailand

Fig. 272: Erection of the rope net

Fig. 273: Installation of the polycarbonate boards

Fig. 274: Lifting in the outer membrane

Fig. 275: Tensioning the outer membrane

Fig. 276: Welding the covering membrane

---

1 Sobek, W.; Linder, J.; Krampen, J. (2004)
2 Holst, S. (2006)
3 Seethaler, M. (2007)
4 Heeg, M. (2007)

3.4.3.4 Erection of ridge and valley surfaces

Ridge and valley surface	Material, structural system	Construction	Equipment
	Membrane material: PES/PVC, Glass/PTFE fabric	Method of stabilisation: stable primary structure	Type of scaffolding: hanging scaffold, abseil working
	Support: at points/corner fittings	Packaging of membrane: longitudinally gathered or rolled	Lifting device: crane
	Edging: rope in sleeve, clamping plates	Load introduction: at points, tensioning/ridge rope	Tensioning tools: hydraulic press, tirfors

Membranes tensioned between ridge and valley can be implemented in radial or parallel versions.

In the parallel variant, the ridge ropes normally connect the high points of two rows of masts, while the valley rope is tensioned from foundation to foundation.

In the radial variant, the primary structure is arranged in a ring complete with the substructure elements. Ridge and valley surfaces are use above all for the roofing of stands in stadiums. The membranes with folded appearance are then mostly tensioned between an outer compression ring and an inner ring cable.

An example for the erection of such a structure is the roofing of the Olympic stadium in Seville (Spain).

After completion of the compression ring and the erection of the ring cable, a temporary net of webbing was installed in the stable primary structure as an area for laying out the membrane. The preparation and the preassembly of the panels and the pulling of the edge rope were done on the ground. Then the longitudinally gathered membrane panels were lifted by crane, laid out on the webbing net and spread out in each direction (Fig. 277).

After the edge elements had been mounted, the ridge ropes were hung. The tensioning of the valley rope against the compression ring introduced pretension into the entire panel (right in Fig. 278).

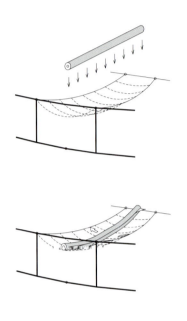

Fig. 277: Erection sequence – ridge and valley surface

Fig. 278: Erection of the roof surface at the Estadio Olympico, Seville, Spain

For the approx. 30,000 m² roof area of the King Fahd stadium in Riyadh (Saudi Arabia), the individual preassembled membrane panels were pulled from the ground with rigging ropes up to the tip of the mast and loaded with approx. 20 % of the intended pretension. After pulling in all panels, the mast heads were pushed upwards to introduce the pretension into the ridge ropes, which pretensioned the system.

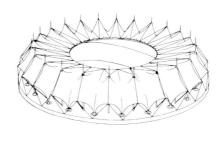

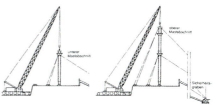

1 Final assembly of the 58 m high masts, stabilisation with temporary ropes.

2 Lifting up the internal ring bundle packet.
3 Pulling in the internal preassembled membrane panels

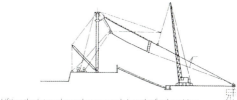

4 Lifting the internal membrane panels into the final position.

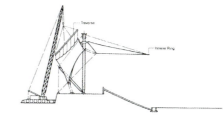

5 Lifting the external membrane panels.

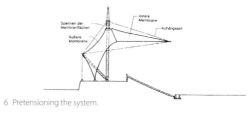

6 Pretensioning the system.

Fig. 279: Roof erection at the stadium in Riyadh, Saudi Arabia

Fig. 280: Scheme of the roof erection of the stadium in Riyadh, Saudi Arabia

A particular form of ridge and valley construction with compartmental pillows was used for the rebuilding of the CargoLifter airship assembly hall to "Tropical Island", a leisure park with tropical rain forest and swimming pools in Brand near Berlin (Fig. 281).

The new requirements for use prompted the owners to rebuild the 360 m long and 220 m wide hall, originally built in 2002, and with a clear height of 99 m, so that the central area of the hall can enjoy daylight as far as possible without impairment and that plants can grow in the hall. To this end, 20,000 m² of the existing approx. 40,000 m² envelope construction – a 2x2 layer PES/PVC fabric – at the south side was replaced by an UV translucent foil covering of 3 layers of ETFE pillow.[1]

After the production of many proposals from various bidders, the decision was made to rebuild the existing ridge and valley construction by installing 14 new diamond-shaped pillows with load-bearing function in each of the 4 central, 35 m wide truss fields (Fig. 282). The 17 x 20 m pillows with a thickness of about 3 m are supported at their outside and inside by form-giving bundles of aluminium-steel ropes. In the valley area at the centre of the field, the rope bundles connect with rope clamps to the existing valley rope, where the edges of the pillow are also held fast by aluminium clamping strips. A thermally insulated GRP rainwater gutter and a condensate gutter are also fixed to this construction (Fig. 283). The fixing of the outer edges of the pillow is to the appropriate upper chord of the 8 m high arch truss of the primary structure.[2]

After the removal of the outer fabric layer of the old envelope, the existing valley rope was shortened, pulled to the correct form with temporary ropes and the rainwater gutter was mounted to it. The existing inner fabric layer was used as a working platform during the erection of the pillow and dismantled and removed after that was complete. The 62 gutter sections per axis with a length of 2.30 m each were fixed to the valley rope in sections working down from the crown of the arch, using a steel component already mounted on the gutter (Fig. 284). At the same time as the erection of the gutter, the surrounding secondary parts were installed, the lower-level rope bundles and the pipes for the computer-controlled inflation air supply.[3]

The transport of the materials to be installed and removed at heights of up to 107 m was a challenge during construction. For this purpose, further mobile access units were installed additionally to the existing provision with hydraulic levelling and rollers, with which the arch construction could be driven along the field edges, and also a remote-control rope traverse was installed in the relevant field (Fig. 285).

The 16 m long membrane packages, delivered by low-loader, were prepared on a scaffold platform, lifted with the rope traverse to the appropriate location for erection and spread on a preprepared webbing net (Fig. 286). Each pillow was prestretched at its relevant installation height, with particular care being paid to the introduction of force into the ETFE material at the weld seams and pillow corners. After the installation of the outer diagonal rope net, the pillow could be supplied with air. The welding of the cover membrane over the gutter then completed the erection of a membrane field.

Fig. 281: Tropical Island, Brand, Germany

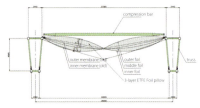
Fig. 282: Section through pillow construction

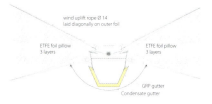
Fig. 283: Section through valley (rope nodes)

Fig. 284: Installation of the GRP gutter

Fig. 285: Access unit and rope traverse

Fig. 286: Installation of the foil pillow

---

1   Blümel, S.; Stimpfe, B.; Rudorf-Witrin, W.; Pasternak, H. (2005)

2   Rudorf-Witrin, W.; Stimpfe, B.; Blümel, S.; Pasternak, H. (2006)

3   Lenk, S. (2005)

3.5 Construction

In order to realistically estimate the problems of constructing a wide-span lightweight structure, it is necessary to possess, in addition to knowledge of the factors resulting from form, force relationships and material, a thorough understanding of the working processes on the construction site. The execution of membrane construction is a sequence of erection procedures. They are central to the practicalities of erection and can be categorised as follows:

Fig. 287: Activities in the implementation of construction

The purpose of erection is to minimise the cost by reducing or simplifying the individual activities. In addition to the effects on production, transport and site conditions, the work planning for the erection process represents the most important factor for optimisation and rationalisation of the construction during the design phase.

3.5.1 Preparation work and preassembly

The joining and completion of single construction elements to large construction units is called preassembly. Before preassembly can be started, the preparatory work must be carried out. The varying manageability of flexible load-bearing elements requires categorisation into

- *Preparation work and preassembly of ropes belong to the primary structure or substructure and*
- *Preparation work and preassembly of membrane panels.*

Fig. 288: *top:* Special transport; *bottom:* Rope drums

Preparation of ropes for preassembly includes the essential steps unloading, temporary storage, unrolling and laying out the ropes. The making of rope nets and the combination into rope bundles and connecting adjacent ropes can be described as preassembly. The working steps necessary for preassembly of membrane panels are the preparation of the membrane joints, edges, corners and high points. Firstly, the membrane must be unloaded, stored temporarily and laid out. Depending on the erection procedure, weight and wind conditions the working level for preassembly is either on the ground or at the erection height.

Further, pretensioning devices have to be set up and measuring equipment installed. Depending on the erection procedure, structural elements of the primary structure may have to be prepared for membrane erection.

3.5.1.1 Transporting, unrolling and laying out ropes

The delivery of ropes is normally in coils, or in the case of larger ropes, on cable drums or reels with the fittings made at the works at both ends. Reels wound in many layers are the most usual type of packaging for high unit weights. They can take 100 m of rope and can weigh 30 t. The delivery is on a special transporter, and the space requirements for lifting devices and for the loading and unloading of the vehicle and the transport on site should be taken into account.

The wound rope coils are pulled off direct from the reel (left in Fig. 289). If the ropes are delivered in loose coils, they must be put on a pay-off stand and pulled off carefully. Reel and pay-off stand should be positioned in the relevant rope axis.

The unrolling process can be carried out under control using hydraulic unrolling devices in order to avoid damage to the rope (see Fig. 290).

Ropes are very sensitive to kinking and twisting. They should be unrolled as far as possible in the air and then slowly low-

Fig. 289: *left:* Rope coils; *middle:* Turning pay-off stand; *right:* Drum reel with horizontal axis

Fig. 290: Unrolling rope from a braked stand

Fig. 291: *left:* Laying out the carrier cable at the stadium in Braga, Portugal; *middle, right:* Preassembly of the cable structure at the Waldstadion in Frankfurt/M., Germany

ered to the ground. Ropes should never by drawn off from coils or reels sideways. In order to avoid damage to ropes, the radius of rope drums should be at least 50 to 100 times the rope diameter.

Mechanical influences through sharp edges, tools etc. can form notches in the rope surface, which can result in later corrosion damage or reduction in strength. In order the lay out ropes in protection, the preassembly area should be cleaned or covered. Laying out areas, which on account of their shape could damage ropes, should be equipped with protective covering and pulleys (left in Fig. 291). If damaged ropes have to be replaced, this can considerably delay the completion date (see section 3.2.2).

3.5.1.2 Preassembly of ropes

After the delivery, temporary storage, unrolling and protected laying out of ropes comes the preassembly. This includes all the necessary working steps of lifting, hanging and pretensioning the ropes. These are essentially the making of rope nets, the making of longitudinal joints, the combination into rope bundles and the connection of adjacent ropes. Also preparation of tensioning devices, like the connection of strand tensioners, or the installation of hydraulic press equipment, can be included in the preassembly. In all this work, attention needs to be paid to the weight of the components to be joined and the selection of appropriate lifting devices.

Fig. 292: *left:* Preassembly of the ring cable at the Gottlieb Daimler stadium; *right:* Preassembly of the hanger ropes at the Waldstadion, Frankfurt/M., Germany

3.5.1.3 Folding, transporting and laying out membrane surfaces

For transport on the construction site, the fabricated membrane is folded or rolled in a suitable manner. In order to protect it from external influences, the packages are mostly packed in cases.

Membrane packages often weigh many tonnes, and therefore attention should be paid to the order and alignment of each package, in order to lift them to the right position and be able to mount them to the primary structure. After investigating the local conditions on the construction site like storage facilities, climatic and wind conditions, the next step is to produce a folding concept in collaboration with the fabricator, which specifies how the membrane is to be folded. A clear description of each panel and of construction groups, which should be marked on each package, is essential.

A folding plan supplied with each package should include details of the centre lines, shape and weight of the individual panels and fields, the description of the edges, seams and corners, and the installation location of each field, the unfolding direction and location of joints (Fig. 293). This can avoid subsequent unnecessary movement of packages, which can often weigh many tonnes. In order to avoid damage, the membrane should be packed with as few folds as possible. When Glass/PTFE fabric is used, which is very susceptible to kinking, the folds should be cushioned in order to avoid breakages in the coating. This could be, for example, soft pipe insulation.

The basic principle should be to create the folding plan with as few folds as possible. In cases where one folding and unfolding process is not sufficient on account of unfavourable transport and unloading conditions on the construction site, then each membrane field will have to be laid out on the ground at the site and folded or rolled again before lifting to the installation location. In order to save working time and cost, this should only be done when it cannot be avoided. Transparent foils can mostly only be folded once.

For unloading and moving the single packages, it is helpful if ropes, chains or slings can be connected.

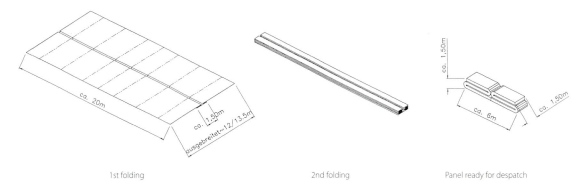

1st folding   2nd folding   Panel ready for despatch

Fig. 293: Folded package for the transport of membrane surfaces

Fig. 294: *left:* Folded package – roofing of the amphitheatre in Nîmes; *right:* Folded package – roofing of the Millennium Dome in London, England

Good accessibility of the prefabricated fittings makes the hanging of temporary ropes of chains easier for lifting.

The storage area for the packages and the laying out area for membranes should be cleaned and free from sharp objects and stones. Laid out foil protects the membrane from dirt. Fittings and zip fastenings are to be specially protected.

Another way of transporting membranes is to roll them at the works. The unrolling on the construction site can, however, be time-consuming. A disadvantage of rolled-up membranes is that they can only be unrolled in one direction and this can only be done where the space is sufficient.

The rolling of membrane surfaces is suitable when the panels can be pulled out in large areas and hung on the stabilised primary structure at erection height. This process is used above all for the erection of arch membranes. An unrolling mechanism mounted on the crane spreader beam is useful in this case (Fig. 296).

Fig. 295: *left:* Roll preparation for the wrapping of the Reichstag; *right:* Roll packaging

Fig. 296: Rolled membrane surface and unrolling mechanism

3.5.1.4 Preassembly of membrane surfaces

Before erecting, lifting or pulling up the membrane into the erection position, various preassembly operations must be done on the panels. This preassembly essentially includes all activities, which serve to join the panel to a larger unit for erection and to prepare the erection. To do this, manual work is carried out on the joints, edges and corners of the membrane panels.

The preassembly work on the membrane panels can be summarised as:

- *Membrane joints – joining the membrane panels (clamping plates, laces, zips)*
- *Membrane edges – Insertion of edge ropes, assembly of edge fittings, preparation of drainage*
- *Membrane corners – Assembly of fittings, connecting keders*
- *Membrane high and low points – Making the edges/connection of clamping plates, ropes, keders*

Preassembly further includes all work necessary before lifting, hanging and pretensioning, like for example stabilisation measures, setting up tensioning aids and devices, laying hoses or installation of measuring equipment. The individual activities are carried out on the ground or at erection height according to the erection procedure.

Fig. 297: Mounting the corner fittings and inserting the edge rope of a four-point awning

Fig. 298: Joining the panels of a circus tent

Fig. 299: *left:* Preassembled high point; *right:* Preassembly of a low point

3.5.2 Lifting and hanging structural elements

After the preassembly is complete, the actual erection of the flexible structural elements starts with the lifting process, the laying or hanging in the intended position. Safety precautions are required to ensure adequate stability of the structural elements and avoid overloading.

3.5.2.1 Lifting and pulling of ropes

To lift and pull wire ropes into position for tensioning, equipment is needed like rope spreaders, pulley blocks, rope winches or strand tensioning devices. The permissible radius of bend should be observed in order not to kink the rope. Ropes should be protected from loading in bending where they come out of fittings. They often need to be protected from damage by twisting.

Ropes are usually lifted into position with rigging ropes. When pulling in ropes, methods from cableway construction are used. The cross-section of the ropes for pulling is often considerably smaller in order to prevent loading of the new rope on lifting.

In the construction of rope trusses, the spanning ropes can be moved and lifted by a remotely controlled trolley mounted on the carrier cable (left in Fig. 301). Another way of lifting and laying ropes at height is to lift the bobbins with a crane and unroll the rope from the bobbin (right in Fig. 301).

In order to lift ropes into position often requires special clamps to be made. Measures also have to be taken on site to turn and anchor the pulling rope.

Fig. 300:
left: Lifting a rope complete with strand tensioning equipment;
right: Lifting a stay rope with rigging rope

Fig. 301:
left: Lifting the spanning ropes at the Waldstadion in Frankfurt/M., Germany;
right: Lifting the ropes at the Stadium in Wolfsburg

Fig. 302:
left: Clamping device for lifting ropes; *middle, right:* Pulley block for pulling rope

Fig. 303: *left:* "Threading in" the columns, *right:* Lifting the rope net for the roofing of the Rhön Clinic in Bad Neustadt, Germany

Lifting rope nets

In order to construct rope nets, the net put together in the works can be laid out on the ground and lifted at points by a mobile crane, and then the net is "threaded in". In the case of the rope net construction in Bad Neustadt, shown in Fig. 303, the net with its approx. 25,000 crossing clips and 25 km rope was mostly installed without stress.[1] The edge columns were leant inwards further while threading in than in their final position, in order to avoid introducing the full stress into the system while lifting. The columns were first tilted to their final position in the last stage of work, which fully pretensioned the structure. Hanging the rope net to the end points largely without stress had the major advantage of being able to work with smaller tools, making the erection quick and cost-effective.

Helicopter erection

Helicopters are also used as transport and erection tools in rare cases. The helicopter is mostly suitable for transport to sites in topographically difficult terrain where access by road is restricted or impossible. The lifting capacity is, however, restricted. The practical working load is up to 5,000 kg. In the construction of cable ways, helicopters are used for mast erection and for transporting the cables, and in power distribution for the erection of high-voltage transmission cables.

Flying erection is risky and can be dangerous. The higher the erection site is (air density) and the warmer the temperature is (thermals), the more problematic it becomes to fly loads by helicopter. Light winds are sufficient to affect erection work.

Considerable experience of erection work is necessary for working with swinging loads. The cost of helicopter lifting is very high and is charged in minutes without the flight to the site. A crew has to be ordered additionally for the maintenance and directing of the helicopter. The preparation work is extensive, and special permission for the flight has to be obtained.

Fig. 304: Rope net erection in Radolfzell, Germany

1 *Inauen, B. (2003)*

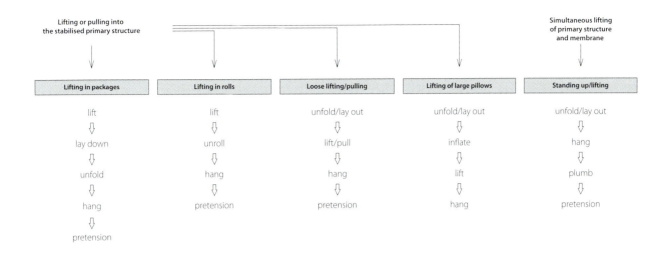

Fig. 305: Lifting of membrane surfaces into position for erection

3.5.2.2 Lifting and pulling of membrane surfaces

In order to install the prepared membrane at the intended location, it has to be lifted to the assembly position. The procedure for lifting depends on the geometry of the structure and stability condition at the time of erection. The method of lifting is also determined by the erection sequence, the size and weight of the unit to be lifted and the height.

The lifting of partial panels from the air into the stabilised structure can be done by lifting folded packages, rolled or loose panels. In many cases membranes are also pulled up along the primary structure or along temporary ropes. Large pillows are mostly inflated at low level and then lifted. Depending to the size of the membrane surface and the weather conditions, the lifting can take many days. If the primary structure is stabilised by the membrane, as is the case with awnings, then the structure must be adequately stabilised during erection and the membrane protected against local stresses.

The lifting of membrane surfaces is one of the most risky phases of erection. The membrane at this stage is normally not tensioned and can be badly damaged by wind gusts. The procedure can therefore only be done when there is no wind or very low wind speeds (light breeze), and this possible delay to erection has to be planned in according to the size and geometry.

In addition to taking wind forces into account, a safety concept has to be considered in case of rain loading. The loading from rainwater produces an increase of weight and can amount to many tonnes of additional loading. If the mem-

Fig. 306: Lifting and unfolding a folded package for the erection for the Airport Center in Munich, Germany

brane is lifted in loose out of the air or pulled up from the ground, then the parts of the membrane, which will be subjected to this additional loading, acute membrane corners, edging, loops or eyes will have to be designed for the loading.

Lifting folded packages

When folded packages are lifted, these are normally put down temporarily on belts, nets or temporary membrane spanned between the members of the primary structure. This method is advantageous because the surface can be unfolded in several directions and no unrolling device is needed.

Lifting rolls

The lifting of rolled membranes is mostly done with a suitable unrolling apparatus. One disadvantage of lifting rolls is that it is only possible to unroll in one direction, which can make the erection procedure more difficult. A spreader beam or rope or belt hanger is normally used, which prevents the kinking of the roll under its dead load.

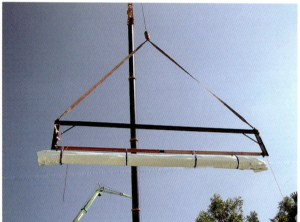

Fig. 307: Lifting of rolled membrane with (left) and without unrolling apparatus (right)

Fig. 308: Lifting of rolled membrane for the wrapping of the Reichstag in Berlin, Germany

Lifting and pulling of loose sheets

When it is not possible on account of the geometry of the structure to lift folded packages or rolls, the partial sheets are lifted in loose or pulled along linear structural elements.

The practicality of lifting or pulling loose sheets depends strongly on the weather conditions. The loose membrane sheets act as sails and are especially at risk of damage. A good weather forecast is needed for the lifting operation. If the edge fittings are mounted on the partial sheets, then the wind conditions are especially critical. Rope spreaders must also be used for lifting.

When pulling in membranes, care should be taken with the strength of acute membrane corners, loops and eyes. Prefabricated fittings make the hanging of rigging ropes or chains for the pulling operation easier.

Large areas, which are pulled up over many days, should always be pulled so that shapes are formed, which allow rain to drain off. If water puddles do form during the pulling operation, they should be cut out or pumped out.

Fig. 309: *left:* Lifting a high point surface; *middle, right:* Lifting the segments for the Forum roof at the Sony Center in Berlin, Germany

Fig. 310: Pulling in the 11,000 m² partial area for the membrane roof over the EXPO grounds in Brisbane, Australia

Fig. 311: *left:* Lifting a membrane premounted to arches; *right:* Lifting a transformable membrane surface

Lifting large pillows

For the roofing of large areas with pneumatically supported pillows, the lifting operation is a major challenge. Many tonnes of fabric and ropes have to be lifted into position and tensioned. This requires precise logistics and technical preparation. The lifting of the pillow into its final position can be done by pulling the loose layers of the pillow along a temporary construction or by lifting in inflated condition.

The parameters determining the erection procedure are the pillow size, the position and construction of the pressure ring, the local conditions, the erection equipment available and the climatic conditions. The pulling of loose large pillows and the lifting in inflated condition are illustrated in the following section by means of three examples.

First erection/roofing of the amphitheatre in Nîmes (France):

To provide a temporary roof over the ancient amphitheatre in Nîmes, it was decided to roof over the elliptical area with the dimensions in the main axes of 62 x 91 m with a light air-supported pillow.

The pillow, held in place by surrounding ropes, will be pretensioned from a compression ring supported by 30 hinged columns. The outer and inner pillow membranes consist of a PVC-coated polyester fabric (upper TYP IV, lower TYP II) and lies on a rope net with rope diameters of d = 22 mm to d = 45 mm. The rise of the upper membrane is 8.2 m, that of the lower 4.2 m. A transparent membrane covers the side festoon.[1]

The entire patterning was calculated by computer. It consists of 500 strips with lengths of up to 84 m and a width of the strips of up to 2.5 m. The approx. 4,000 m² upper membrane

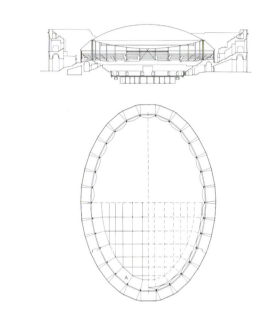

Fig. 312: Aerial view, section and plan of the amphitheatre in Nîmes, France

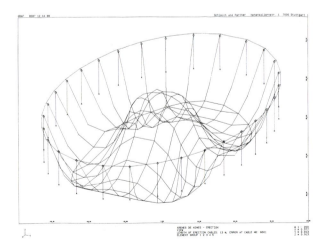

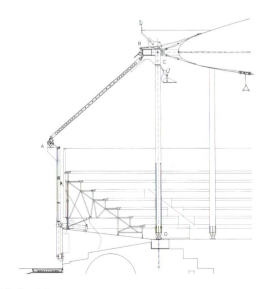

Fig. 313: *left:* Scheme for the lifting procedure of the upper and lower membranes; *right:* Sectional view

---

1 Habermann, K. J.; Schittich, C. (1994)

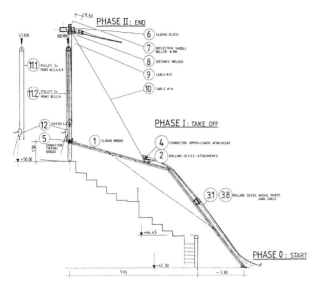

Erection sequence:

Erection of the columns and the compression ring.

Erection of the bridges and the temporary construction.

Preassembly of the rope net groups.

Laying out the rope net and pulling up to the kink.

Unfolding of the lower membrane.

Unfolding of the upper membrane

Preparation of the membrane layers to pull in.

Pulling up of the package with tirfors and tackles along the temporary bridge up to the kink at a height of about 8 m, erection there of the rope net and ring rope fittings.

Pulling the entire package to within 50 cm of final position.

Fixing and rehanging on hydraulic pulleys.

Pulling to final position at the compression ring.

Fixing the fittings to the columns.

Connecting the blower and inflation of the pillow.

Fig. 314: *left:* Slide bridge; *right:* Erection procedure

and the lower membrane of nearly the same size were fabricated in one piece each.[1] The pillow can be walked on for maintenance.

The challenge of the initial erection was to lift the outer and inner layers of pillow and the rope net into position for tensioning without damaging the historic stands.

The first scheme, to lift both pillow layers by inflating a pillow construction underneath (left in Fig. 313), was investigated and rejected.[2]

The erection procedure finally used was to pull the supporting rope network and the loose layers of pillow in stages along 30 steel slideways mounted on the masts, tension

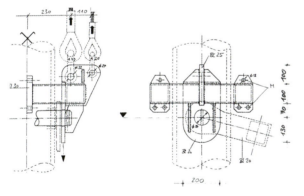

Fig. 315: Temporary construction to pull the pillow along the masts

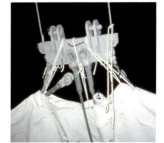

Fig. 316: Pulling the pillow along the slideway bridges

---

1 Bergermann, R.; Sobek, W. (1992) *2 Dürr, H. (2003)*

Fig. 317: Initial erection of the pillow over the amphitheatre Arénes de Nîmes, France

1 Erection of the temporary bridge
2 Laying out the rope net and the lower membrane
3 Laying out the upper membrane
4 Pulling up Phase 1
5 Detail / festoon apex
6 Rehanging, pulling up Phase 2
7 Spanning the festoon ropes to the compression ring
8 Connecting the blower units
9 Inflated pillow with festoon membrane
10 Dismantling the temporary bridge

them to the perimeter compression ring and inflate the pillow by introducing compressed air. The illustrations in Figs. 314 to 316 show the sliding bridge construction, the erection phases and the equipment for the lifting process.

The size of the package of membrane to be unfolded was 10 x 7 x 2.5 m with a weight of about 8 t per package. The weight of the entire membrane package including rope net was approx. 40 t. Extensive temporary construction measures were necessary for the lifting process. The erection of the pillow lasted approx. 2 weeks.[1]

In order to remove the pillow again in the summer months, an economic scheme for the yearly erection and dismantling of the entire construction had to be developed.

This scheme provides the columns with a hydraulic raising and lowering mechanism. To lift the pillow layers complete with rope net, stayed lattice booms are fitted to the top of the columns. The forward corners of a triangle formed of column, distance beam and telescopic boom form the support for the boom (Fig. 318, Fig. 319). The telescopic beams are adjustable for length with two bolts in order to be able to adjust the resting point of the boom within millimetres: this should ensure that at the end of the lifting process (on reaching the compression ring), the bolt holes and the nodes of the compression ring exactly line up with the bolt holes of the pillow fittings.[2]

The flexible edge detail of the pillow with 30 festoon ropes per pillow and the bundling of the erection connections at the column heads make the erection and dismantling work easier. The process of lifting the pillow lasts about 5 hours. The dismantling is done in the opposite order.

The erection procedure for lifting the pillow parts is summarised in Fig. 319.

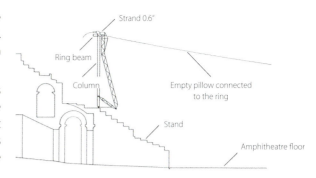

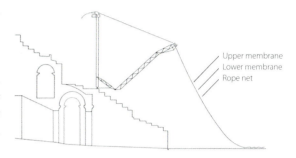

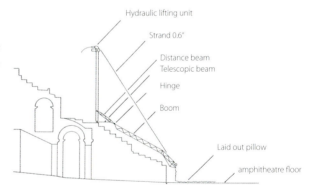

Erection procedure::

Putting up and staying every second of the 30 columns.

Lift the compression ring with the lifting equipment installed on the column.

Connect the ring with the columns.

Erection of the remaining 15 columns.

Installation of the boom beam.

Assembly of the blower equipment.

Lay out the rope net and the lower and upper membranes on the amphitheatre floor.

Close the closing membrane, mount the festoon rope, fix the pillows to the boom beams.

Lift the pillow.

Fix the pillow to the ring.

Inflate the pillow.

Install the façade and remove the boom beams.

Fig. 318: Distance beam, telescopic beam and lattice boom

Fig. 319: Scheme of lifting process

1 Dürr, H. (1988)

2 Bergermann, R.; Sobek, W. (1992)

a Lay out the pillow pieces and lower the booms

b Connect the pillow pieces to the booms

c About 1/3 of the lifting travel

d About 2/3 of the lifting travel

e Pillow pieces without ground contact

f Condition immediately after inflation

Fig. 320: Annual process of lifting the pillow

Mobile roofing for the Arena in Vista Alegre:

An example for the lifting of a pillow in inflated condition is the large pillow erection for the roofing of the bullfight arena in Vista Alegre near Madrid (Spain). A convertible pillow constructed on the existing roof structure can be opened within 5 minutes in order to ventilate the arena with more than 14,000 seats.

The pillow with a diameter of 50 m consists of a compression ring of parallelogram section, an outer translucent PVC polyester membrane and an inner ETFE foil supported by a rope net (Fig. 323). At the thickest point, the membranes are 12 m apart.

The opening for the pillow in the fixed gridshell roof construction is formed by the parallelogram-section compression ring, on which 12 stayed 12 m high rocker columns stand. The square 300 x 300 x 10 mm hollow profile columns, which also act as horizontal bearings for the pillow under wind loading, are connected at the heads with ring ropes (Fig. 324). The 60 t pillow has a plan area of 1,960 m² and is raised and lowered by 12 rope winches situated at the foot of the columns over rope wheels on the heads of the columns (Fig. 321). It can be lifted 10 m and can be parked in various positions.[1]

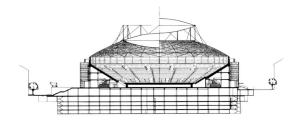

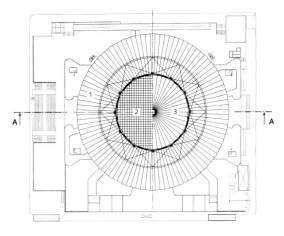

Fig. 322: Plan and section of the Arena in Vista Alegre, Spain

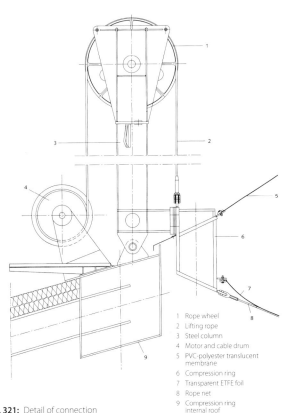

Fig. 321: Detail of connection

1 Rope wheel
2 Lifting rope
3 Steel column
4 Motor and cable drum
5 PVC-polyester translucent membrane
6 Compression ring
7 Transparent ETFE foil
8 Rope net
9 Compression ring internal roof

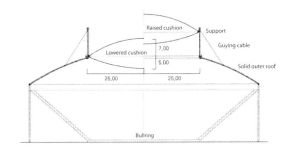

Fig. 323: Scheme of the pillow roofing

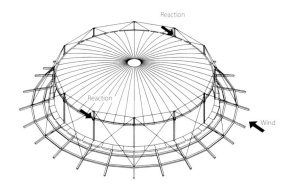

Fig. 324: Isometric

---

1 Schlaich, M. (2000)

To lift the pillow into its final position, it has to be raised 30 m in height. The play was only a few millimetres. This requirement meant that the columns had to be aligned exactly through the rope stays, in order that the pillow guide would not jam on the steel columns.[1]

Erection procedure::

1 Standing up the rope-stayed columns on the fixed outer roof. Erection of the pillow compression ring made of prefabricated sections on a scaffolding in 5 m height.

2 Installation of the upper membrane and stabilisation through temporary ropes to the column feet.

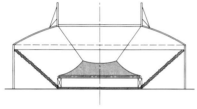

3 Installation of the lower foil and hanging of the rope net onto the pillow compression ring.

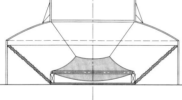

4 Inflation of the pillow and checking the function of the blower and the air-tightness of the pillow.

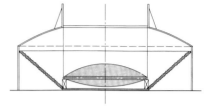

5 Lifting the pillow with temporary ropes into its final position.

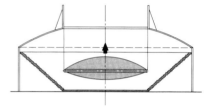

6 Fixing the bearing brackets to the pillow compression ring and fixing the rain protection plates.

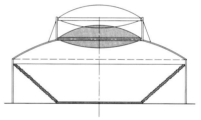

Fig. 325: Scheme of lifting process

Fig. 326: Lifting the pillow

1 Schlaich, M. (2000)

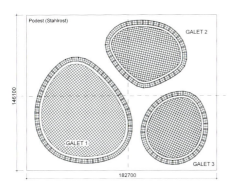
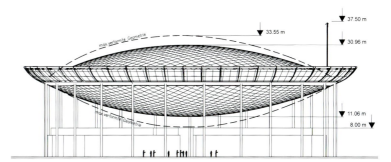

Fig. 327: *left:* Plan of Galets No. 1 – 3; *right:* Section of Galet No. 1; Neuchâtel, Switzerland

Erection of the Galets for the EXPO.02:

An example for the open-air lifting of pillows is the construction of the Galets, three large pneumatic pillow structures, which were built as roofing for the exhibition, part of the Swiss national exhibition EXPO.02 at the Arteplage in Neuchâtel.

The rims of the two-layer, lense-shaped pillows with diameters of 61 – 92 m are anchored to a steel perimeter compression ring at a height of about 15 – 21 m. The box-section compression ring is supported by steel columns, which are supported on hinged joints onto piles rammed into the lake. [1]

A cantilevering membrane fixed to an edge tube forms the upper edge of the structure and clads cantilevers welded to the compression rings (left in Fig. 329).

Because of the dimensions of the three large pillows, which cover an area of 3,500, 4,000 and 9,000 m² respectively, the PES/PVC membranes had to be made in 2 pieces for the Galets No. 2 and No. 3 and in 4 pieces for the Galet No. 1 and joined together on site with assembly joints. The welded strip widths of the panels were between 1.8 m and 2.3 m.

Fig. 328: Galets on the Arteplage in Neuchâtel, Switzerland

In order to enable a flat underside, the pillows were built without supporting rope nets. To achieve this, the internal pressure had to be limited to about 0.25 – 0.4 kN/m² according to the weather situation in order to avoid damage to the fabric. Because the design loads could not be resisted with such a low internal pressure, the Galets had to be equipped with a gas-fired heating system in order to combat excessive snow loading with warm compressed air. The heating mounted on the compression ring of the Galets was controlled auto-

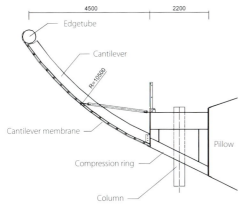

Fig. 329: *left:* Section through the upper ring; *right:* Box-section beam with cantilevers

[1] Bukor, S. (2003)

Fig. 330: *left:* Preassembled compression ring; *middle:* Lifting the columns; *right:* Preassembled pillow

matically by a weather station installed in each Galet, which could measure wind speed, air pressure and air temperature and detect snow. For the security of the structures, twice the necessary number of blower units were installed and connected to an emergency power supply, in order to supply the three Galets with sufficient electricity in case of power failure.[1]

For the advance assembly of the three pillows complete with steel construction, it was possible to use the exhibition platform. After ramming the piles, the prefabricated sections of the steel construction were put on trestles and welded. The trestles were arranged so that they could stabilise the columns for the preasssembly and lifting processes.[2]

Then the columns were lowered though the 1.2 m x 1.2 m openings in the compression ring one after another by a truck crane. After being marked, the membrane panels were rolled out, unfolded and prepared for assembly. The connection of the clamping plate joints from the middle outwards, the individual panels of the lower and upper membranes were joined to each other. Then first the upper and then the lower membranes were clamped to the perimeter compression ring from the panel joints outwards (Fig. 331).[3] To do this for Galet No.1 required about 150 tirfors, which were distributed around the ring.

After the inflation of the pillows and the installation of all electromechanical equipment, the compression ring together with the connected pillow could be lifted into the intended position using hydraulic presses. For Galet No.1, this meant equipping every third column with lifting equipment, each with 2 hydraulic presses, in order to lift the construction with a weight of 4,700 kN (Fig. 332). During the lifting process, 2 columns at a time were stabilised by temporary masts standing outside the Galet, which were connected to the columns by horizontal struts.[4]

When the correct height was reached, the compression ring was welded to the columns and the columns fitted with cross-bracing ropes.

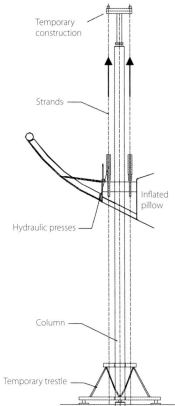

Fig. 331: Assembly joints Galet No. 1

Fig. 332: Lifting mechanism

1 *Ryser, R.; Badoux, J.-C. (2002)*
2 *Ryser, R.; Badoux, J.-C. (2002)*
3 *Dürr, H. (2002)*
4 *Ryser, R.; Badoux, J.-C. (2002)*

Fig. 333:
left: Erection of the clearance hall at the passenger quay in Warnemünde, Germany; *right:* Assembly of a stage roof in Travemünde, Germany

Simultaneous erection of primary structure and membrane

In order to erect membrane structures, which only have sufficient stability after completion, measures have to be taken to ensure the structural safety of the structure and the stability of the elements of the structure at all phases and conditions (see section 3.4.1.2). The erection of such a structure is mostly done by pivoting or lifting the primary structure complete with the already connected membranes and ropes (Fig. 333).

Extensive preparatory works are needed before this can be done. The membrane must be attached to the stiff load-bearing elements of the primary structure so that the fabric material and the ropes are kept free of damage at every stage of the process. The freedom of movement must be checked in every direction.

If the primary structure is to be lifted complete with the membrane panels attached to it, then the capability of the membrane corners to resist this loading case should be checked (left in Fig. 333). It is also important to avoid kinking of the edge rope and the foot of the column jamming in the joint. Temporary situations should be controlled with rigging ropes, which can be removed rapidly. The erection should take place when there is no wind.

When the structure is pivoted up into position, the stiff structural elements have to be positioned in their starting position so that they can be pivoted exactly into the bearing plane. Eye plates mounted on the foot of the mast should not be deformed. If masts are to be stood up by pivoting, they should be restrained against sideways movement. The example shown in Fig. 334 shows the process for the erection of a Glass/PTFE fabric for a market roof at Zeltweg, Austria. An awning was erected there by diagonally staggered pivoting.

a Initial situation with preassembled membrane

b Pivoting of the first mast

c Diagonally staggered pivoting

d Interior view with temporary stays

Fig. 334: Erection of the roof over the marketplace in Zeltweg, Austria

3.5.3 Introduction of loads – pretensioning

The load-bearing behaviour of wide-span, lightweight structures is determined to a great extent by the pretension and its interaction with geometry and material. The method of introducing the pretensioning forces is of central importance in the implementation of the erection.

Purpose and function of pretensioning

Flexible load-bearing elements in form-active structural systems have to be tensioned in a tangential direction in order to achieve sufficient stiffness of the system under external loading. The tensioning of the structural elements is intended to act against any deformation of the structure and maintain its intended form. In order to achieve the necessary form stability of the system, tension forces have to be introduced before these forces act. Flat or spatial structures or elements, which are under tension in a weightless condition, are therefore described as *standing under pretension*.[1]

The introduction of the pretensioning forces creates an internal stress condition, which stabilises the structure under serviceability loading. The level of forces to be introduced to achieve the pretensioned state must be in equilibrium with the expected loading. The stress distribution must be so that an equilibrium shape is taken up by the system. The dependence of the loading on the geometry of the surface, expressed by the curvature, arises from these deformation conditions for form-active systems – form and stress distribution determine each other.

Flat tensioned surfaces and structural elements have insufficient stiffness under loading. The deformation under loading is large, and no equilibrium is possible. Only with increasing curvature of the surface or with increasing angle between the rope axes can sufficient stiffness build up. Increasing curvature leads to reduced support reactions and the influence of the level of pretension becomes less (left in Fig. 335). If the geometry of a membrane surface has a positive Gaussian curvature, it is called a synclastic surface; with curvature in opposite directions, antisynclastic. For anticlastic membrane surfaces in a pretensioned state without external loading (q = 0), a fixed, no longer arbitrary relationship ensues for both curvature directions (right in Fig. 335). They form, because of their geometry, a higher initial resistance to loading then a surface curved in the same directions.

In addition to the geometrical stiffness, the extent of the surface curvature of the surface elements also directly influences the construction of the foundations, which transfer the forces into the subsoil. It is therefore a good idea to choose a form from the multitude of possibilities, which does not result in unnecessarily high pretensioning forces to limit the deformation, and also transfer the forces into the subsoil through a clever combination of compression and tension foundations.[2] This can balance forces and save costs.

The reordering of the internal forces and stresses caused by the pretension produces a favourable alteration of the statical system with regard to reducing the deformations and an improvement of the stability relationships. Another important aspect of the application of pretension is that the cross-section is fully exploited by tension loading. The load-bearing elements, which are predominantly loaded in tension, can have considerably more slender dimensions than elements under, for example, bending, leading to a substantial reduction of the weight of the structure.

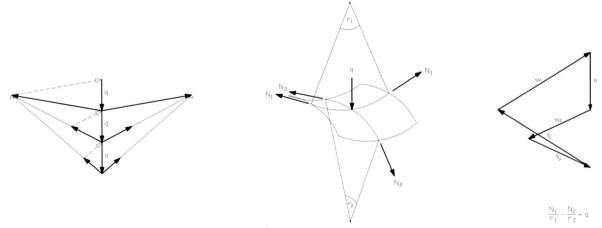

Fig. 335: Equilibrium models: Increasing resistance to loading through alterations of the angle of the system axes to each other

---

1 *Palkowski, S. (1990)* 2 *Leonhardt, F.; Schlaich, J. (1973)*

Application of pretension to lightweight structures

Because of the negligible bending stiffness of the sheet and linear elements and the relatively very light mass of lightweight structures, large deformations and oscillations can be caused by wind loading, which can influence the structure dangerously. The pretensioning of the flexible load-bearing elements between fixed points, edges or corners increases the shear stiffness of the individual structural elements and the stability of the entire structure, which resists against the danger of this deformation. Through the pretension, the shear loading is transferred so that the load-transferring flexible load-bearing elements are predominantly loaded in tension and the necessary stiffness is preserved under load.

It should be mentioned that on principle, when slender structural components and large spans are desired, load-bearing elements, which are stiff in bending or compression, like columns, pylons and edge beams, can also be prestressed. Depending on the material, geometry of the element, weight and load-bearing behaviour, various processes are used, with and without tendons. The most suitable materials for the introduction of prestressing forces are reinforced concrete and steel. The production of prestressed load-bearing elements out of timber is difficult on account of the shrinkage characteristics of the material and is still being researched at the moment.

The known methods of pretensioning for flexible load-bearing elements are the mechanical pretensioning of ropes, webbing, fabrics and foils and the pneumatic pretensioning of fabrics and foils. Another way of achieving form stability is pretensioning through loading, as is the case with hanging roofs under ballast, where the weight is higher than the wind uplift to be expected. Pretensioning with dynamic forces, as has been investigated for rotating umbrellas[1], is of no use in construction at the moment because of the lack of form stability.

Tensioning process

The gradual build-up of the state of pretension until the intended geometry is reached is described as the *pretensioning procedure.* It is the result of considering all structural and construction considerations, whose purpose is to achieve a pretensioned structure as end result.

The most important criteria for the selection of the pretensioning procedure can be described as shown in in Fig. 336:

During the tensioning process, construction elements can be heavily loaded. The structural element to be tensioned, its adjacent substructure elements and the entire structure are sometimes subject to large deformation during the introduction of the forces. The controlled introduction of tension force, the direction and sequence of tensioning therefore need to be very carefully planned. The stability of the structure and its elements during the tensioning process should not be endangered and the tensioned element should not be overloaded. The tensioning stages and tensioning direction are determined through the investigation of the deformations, stability investigations, structural calculations for the entire structure as well as its parts for each erection condition and the appropriate pretensioning procedure is selected. The tensioning concept worked out determines the necessary control and compensation measures and defines the tolerance ranges.

The most important tensioning procedures for introducing the tensioning forces into flexible structural elements are illustrated and described in the following section.

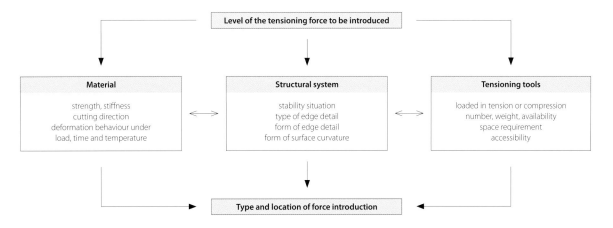

Fig. 336: Criteria for the selection of tensioning procedure

1 *Baumüller, D. (2000)*

Tensioning direction for flexible surface elements depending on material

The stretch properties of the material used, its coating and its cutting pattern are of considerable importance for the tensioning procedure (see section 3.4). The materials most commonly used are polyester/PVC fabric and PTFE/glass fabric with dissimilar stretch properties in warp and weft direction and polyester/PVC fabrics with the same stretch properties and fluoropolymer foils.

Fabrics with dissimilar stretch properties have a higher stiffness in the longitudinal fabric direction than in weft direction; the stiffer warp direction stretches very little under load. Care therefore needs to be taken in the patterning that the calculated share of compensation is considerably less in the stiff warp direction than in the more flexible weft direction. This situation can be used to advantage for the tensioning procedure.

When the curvature conditions are favourable, the distribution of stress in the fabric is ideal and the edge is stiff in bending, tensioning in one direction alone can tension the other direction as well. This means that the fabric threads in the transverse direction shorten as well, and the membrane is also tensioned in this direction.[1] If the edge detail features clamping plates, it is important that the membranes should have their final length before the assembly of the edge fittings along the edges. The shear stiffness of the coating is significant when tensioning the edges. Especially when PTFE-coated glass fibre fabrics are used, particular care and attention need to be taken in the sensitive edge areas.

When the geometry of the surface has equal dimensions in the anisotropy directions, it can be a good idea to specify fabric with similar stretch properties in the warp and weft directions. The nearly equal thread lengths can be favourable for the tensioning process in that the equal tensioning travel distances can be brought about with the same force. Because the weft thread curvature of this fabric practically corresponds to the warp thread curvature and the coating thickness on the ridge of the thread is considerable greater than with conventional fabrics of the same weight per area, a higher force level is nonetheless required for tensioning. Such fabrics are suitable above all when the geometry has a less curved surface.

The construction of anticlastic surfaces with single-layer, isotropic fluoropolymer foils is limited to a few metres of span. These foils can only be tensioned by the application of linear tension through stiff edgings. Foils are often used in multi-layer pillows. These are often designed with rope net support. Because of the relaxation properties and the creep behaviour of flat foils, such surfaces normally have to be retensioned.

In the following section, the main emphasis is on the tensioning of fabrics with dissimilar stretch properties on account of the frequency of their use.

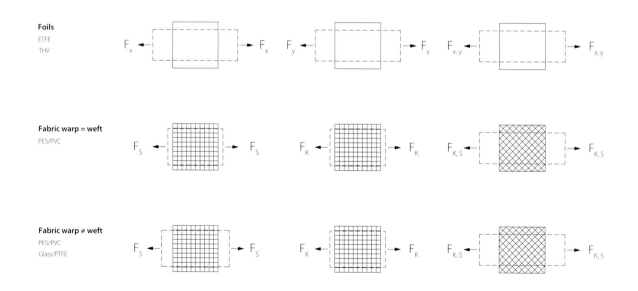

Fig. 337: Scheme of the elasticity behaviour of membrane materials with similar and dissimilar stretch properties

---

1 Basic statements about the influence of cutting patterns on the tensioning of fabrics with dissimilar stretch properties have been made in section 2.4.3.1.

Level of tensioning force to be applied, stress distribution and tensioning sequence

In addition to the direction of force introduction, the level of tension force to be applied and the sequence of the tensioning stages are also of central importance for successful erection.

To achieve the design stresses, the compensation must be fully pulled out of the fabric. Two preconditions for this are the correct level of tensioning force applied and the observation of the specified tolerances of the primary construction. The stretching travel length is calculated from the calculated compensation values, taking into consideration the enlargement when load is applied and the relaxation behaviour with load, time and temperature.

Depending on the material, geometry, cutting pattern and tensioning direction, areas of the surface without stress can occur during the tensioning of fabrics and foils, which spread over the surface through force diversion. In order to achieve the calculated stress distribution in the surface, the fabric has to be overstressed in certain areas. The tensioning equipment cannot be laid out for just the tensioning force calculated for the equilibrium state, but in some cases has to be capable of many times more. If the fabric is overstressed, this must not exceed the relevant breaking stress of the material, which could lead to failure of the membrane.

In order to keep this overstressing as low as possible, and so as not to cause additional relaxation through overloading, the tensioning process must be done slowly. The distribution of stress during the tensioning process can, depending on the material, extend over a long time. The single tensioning steps for PES/ PVC membranes extend over many hours, and for Glass/PTFE fabrics, this can be several days on account of the low stretch of the glass fibres. The temperature also has an effect on the stress distribution in fabric. No PTFE membranes should be installed under 8 °C.

If it is not possible to tension slowly, and the fabric or foil has to be installed quickly, then the biaxial tests have to be performed according to the real situation. This means that a large strain is applied and the relaxation and the residual pretension are determined. If this does not produce the required pretension after the first test, then the test is carried out iteratively. [1]

The decision which element is tensioned, when, with what force and in which direction is often based on practical considerations. The principle for erection is how and at what stage is it possible to tension efficiently with the least effort.

The effects on the material and on the load-bearing behaviour in the temporary situations during erections should therefore be considered at an early stage of the design process.

Recording of all tensioning stages during the implementation of the planned pretension is a precondition for calculable load transfer and a membrane without folds.

Consideration of retensioning equipment

To estimate the risk of wrinkles, the relationship of relaxation behaviour and applied load must be investigated. The relationship between relaxation travel and geometrical stretch to material strength should be evaluated and, if necessary, methods of retensioning the membrane surface should be considered.

Fabric membranes with flexible edges can generally be retensioned by shortening the edge rope or by readjustment of the corner fittings. If the fabric is spanned between rigid edges, then later corrections to the pretension can be carried out by moving the edge fitting. If this is not possible, and if assumed elastic properties of the material cause considerable variations from the calculated section values, then *"… the achievable precision of the tensioning forces depends directly on the precision of the forecast alteration of the length of the membrane during pretensioning."* [2] Extensive (biaxial) tests on the material and meticulous measurements are then unavoidable.

Retensioning of foils can, depending on the load behaviour, make a longer extension necessary than for membranes. If the foil is mechanically tensioned, this can be reacted to by the installation of compression springs. Pneumatically supported constructions are normally built without a mechanism for retensioning. Retensioning equipment like, for example, a spring mechanism, would be very expensive to produce and install for a compartmental pillow construction. Retensioning is usually done by controlling the inflation pressure.

---

1 Stimpfle, B. (2004-2)

2 Baumann, Th. (2002)

Types of pretension application

In order to make a fabricated membrane sufficiently load-bearing in bearing and tensioning directions, it has to be tensioned and thus enlarged. To achieve the intended geometry, the stretch distance from the compensated membrane edge to reaching the anchoring point has to be accomplished. This tensioning travel can be overcome by applying edge loads tangentially into the surface or by nontangential application of the load.

When the tensioning is done tangentially, the geometry and stiffness of the edge detail determine whether the loads are applied linearly or at points. If the membrane has an edge, which is stiff in bending, then the loads are applied with a linear distribution. If the edge is flexible and curved, then the pretension is applied at points. The application of loads to surfaces can be by creating pressure differences or by mechanical loading.

The following section illustrates various principles and procedures for introducing the pretension into membrane surfaces, from directly pulling the edges or corners of the membrane element with tensioning aids to alteration of the position or shape of the primary structure or the substructure (Fig. 338). The diagrams are to be understood as sketches showing the principles. In the practice, many of the principles explained are combined during erection.

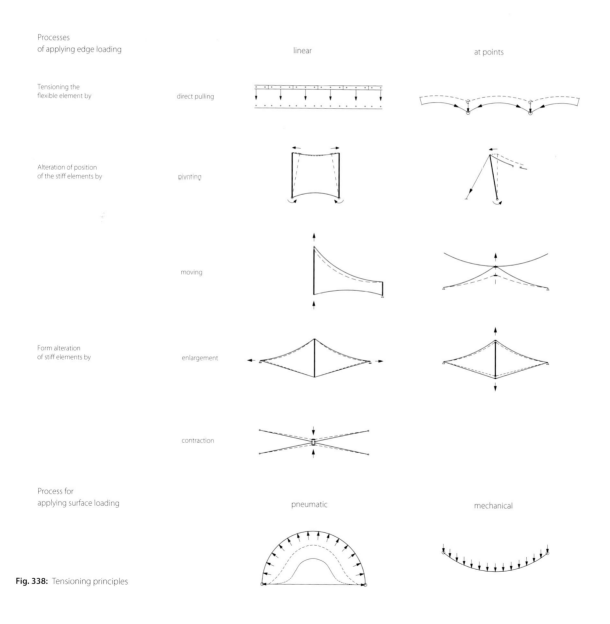

Fig. 338: Tensioning principles

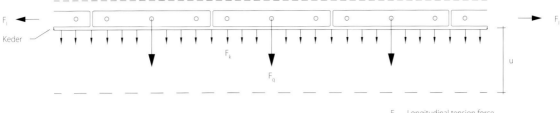

Fig. 339: Linear application of edge loads through a stiff clamping plate edging

F_l Longitudinal tension force
F_q Transverse tension force
F_k Linear force distribution
u Tensioning travel

3.5.3.1 Linear application of edge loads

The tensioning of membrane surfaces with stiff edge details is done by the application of linearly distributed loads. The load transfer is through tensioning tools, mostly ratchet tie-downs, tirfors or hydraulic presses, through the edge element linearly onto the keder (Fk in Fig. 339). The smaller the separation of the points where the load is applied is, the more even is the distribution of forces and stresses in the membrane. Depending to the layout of the clamping plates and edge curvature, a decision has to be made how often the force has to be applied over the length of the edge to distribute the loads evenly onto the keder.

The tensioning of the stiff edge is done in many stages depending on type of fabric, type of material, cutting pattern and form of edging. The main pretension is applied gradually and then the surface is finely tensioned. When the geometry of the surface, the type of material or the climatic conditions are unfavourable, and losses of pretension can be expected during staged tensioning, compensation will also be required at many intermediate stages. Various pieces of equipment are used in the tensioning process, which either grip the keder directly or are installed between the edge fittings and the mechanical or hydraulic tensioning tools (see section 3.3.3). The surface can be finely tensioned or retensioned using permanently installed tensioning equipment or movable edge elements.

Direction, travel and sequence of tensioning

An important factor in the creation of the tensioning scheme for fabric with a stiff edge detail is the direction and sequence of the forces to be introduced depending on the location of the anisotropy axes. It is of central importance for the tensioning process during erection when force acts in one direction, how the forces and strains in the other direction are influenced. In addition to the influence of the transverse strain and the shear stiffness in the fabric, the different stiffnesses of clamping plates and fabric and the resistance to the rotation at the stiff edge are decisive and require a meticulous procedure for tensioning. In order to avoid kinking of the clamped area relative to the force direction, attention must therefore be paid especially to tangential introduction of load when the edge detail is stiff.

Different forces are to be applied to pull the membrane into position depending on the length of the tensioning travel. The level of tensioning force to be applied has an effect on the equipment to be used for erection and the working time. During erection, higher forces than the forces, which will act on this area in the finished state, can be applied at points. In order to determine the primary tensioning direction, it should therefore be considered what level of force is required for the pulling of the tensioning travel. The basic question is, from which direction is it easier to pull the compensation out of the fabric? In addition to the extent of the forces to be applied, the spatial conditions are also an important factor in this decision. Sufficient room for the fixing of the tensioning equipment is just as necessary as the question which structural element is to restrain the tensioning. The stability of such elements has to be checked.

Fabrics with dissimilar stretch properties are fabricated with the two anisotropy directions differently shortened. This fact can make the tensioning of such fabric easier. In the ideal case, the less compensated weft direction can be completely tensioned just by pulling the fabric in the warp direction. If this favourable circumstance for erection cannot be exploited, then it will have to be pulled in both directions. For this reason, care is taken in patterning that the compensation parallel to the stiff edge is as low as possible. If the geometry and the use of materials with dissimilar stretch in warp and weft direction has the effect that no interaction is possible between warp and weft directions is possible, then the layout of the strips will have to be altered.

A typical example for the application of this principle of linearly distributed introduction of edge loads is the erection of membrane surfaces spanned between two arches (Fig. 340).

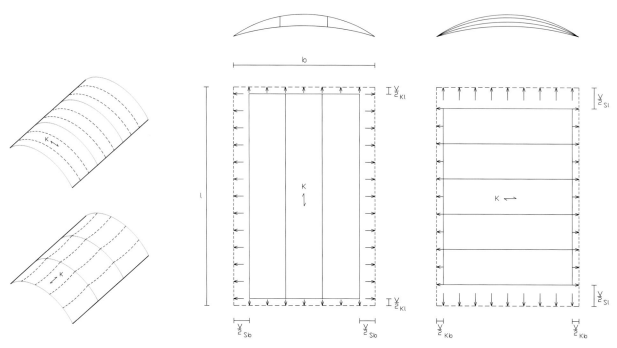

Fig. 340: Tensioning travel for different cutting pattern directions of an arch membrane with stiff edging

The unrolling of the membrane perpendicular to the arch usually results in the least distortion for the pattern. The warp direction then runs in the plane of the arch and the tensioning is done in the weft direction. If it is not possible to tension in this direction (for example because of restricted space), then the cutting pattern will have to be turned. The cutting direction is, however, normally laid over the shorter side of the surface.[1]

After lifting, the fabric is clamped to one arch then pulled to the opposite arch and successively tensioned and clamped from the middle to the outside.[2]

It is advisable for many structural geometries to tension the weft direction over a longer travel distance with less force. The considerably stiffer warp direction should, on the other hand, be tensioned over as short a travel as possible. It can be sensible for the practicalities of installation not to compensate in warp direction and only pull the constructive stretch out of the material. In order to avoid sideways drifting of the forces when tensioning over long travel distances, the stretch should be introduced as evenly as possible starting from the middle. If there is not enough space to do this and the membrane has to be tensioned from one side over the entire length, then correspondingly powerful tensioning equipment will be needed.

In every case, the space requirements for installation of the tensioning equipment should be taken into account in the design of the edge detail.

If insufficient space is available between the adjacent keder rails or clamping plates of two converging stiff edges (left in Fig. 341), it will not be possible to tension the fabric start-

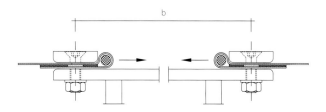

Fig. 341: Space requirements for the tensioning process

1 Essrich, R. (2004)
2 Böhmer, C. (2004)

Fig. 342: Tensioning a clamping plate edge (PES/PVC fabric)

ing from the middle. In such cases, another method of introducing the tension will have to be found.

The situation is different with clamping plate joints, which are preassembled without tension, or edge details for two-layer pillow constructions, where the full pretension is first applied with the inflation (right in Fig. 341). When constructing between pillows, sufficient room for tensioning will be required.

The calculated shortening of the considerably more strongly compensated weft direction can be pulled out of the membrane surface parallel or perpendicular to the stiff edge. If this is not possible and the membrane has to be tensioned in both weave directions in order to obtain an end result without wrinkles, then measures are taken to enable the membrane surface to span in the long direction. Depending on the material, edge detail and strip layout, the surface will then be tensioned in the long direction simultaneously or after the transverse tensioning.

The edge of the fabric to be fitted into a clamping plate can be prestretched parallel to the edge before bolting the clamping plates. This is either done manually by prestretching a piece at a time in the long direction or by applying presses over the length of the edge. Another possibility is to first pull the surface in the transverse direction and then tension in the long direction. This is done by hanging the keder in the lower fixed half of the clamping plate, loosely mounting the upper half and tensioning in the long direction.

If the stiff edge is assembled without tensioning the fabric parallel to it, then the calculated compensation value can

Fig. 343: Transverse tensioning of a keder rail edge (Glass/PTFE fabric)

Fig. 344: Longitudinal tensioning of a keder rail edge (Glass/PTFE fabric)

only apportion to the spacings between the clamping plates when the pretensioning force is introduced. The fabric could tear at this point. When tensioning edges with one-part keder rail in the long direction, it is advisable to mark the membrane edge and keder rails, in order to pull the membrane edge to the correct position. To improve the sliding in the keder rail, lubricant or talcum powder can be used. The keder seam should be tidily welded.

The subsequent tensioning in the long direction can cause problems if the profile is too tight or the membrane comes out of the profile kinked. Keder cords with insufficient Shore hardness can prevent smooth sliding. Pulling lugs with built-in keders can be used here to assist longitudinal tensioning.

When temporary membranes are spanned between stiff edgings and the edge elements are to remain visible, the edges can be constructed as shown in the following section. Quick changing of the membrane is then made possible by the easily accessible tensioning elements.

In a tube edge detail, a steel tube pushed through the sleeve functions as a keder. The load is introduced by even pulling

Fig. 345: Stiff edge detail at the Graz airport tower in Austria

over the entire length of the tube. A suitable anchorage is standing bolts welded to the primary structure. Tube edges can be retensioned using threaded tension rods.

When tensioning clamping plate edges, which are mounted on the edge rope with sheet metal loops, the ability of these to move along the rope is a problem. Care should be taken that the sheet metal loops do not slew, jam or damage the rope when the load is introduced.

Fig. 346: *left:* Tensioning a tube edge; *right:* Tube edge against standing bolts

Fig. 347: Tensioning a clamping clip edge

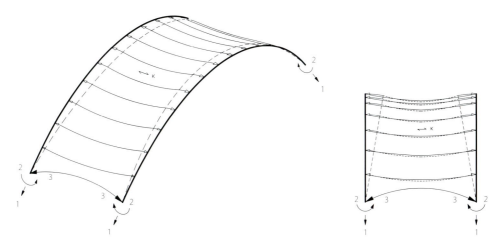

Fig. 348: Introduction of tension by tilting the edge beam

Application of loads by tilting the edge beams
The introduction of linear edge loads into the membrane surface can also be achieved by changing the position of the stiff edge beam. One example of this is the tilting of arch-shaped edge beams. This is an efficient method of tensioning, which is mostly used with many parallel rows of edge arches. The membrane surface is prestretched in the arch direction and clamped to the arches (1 in Fig. 348). After pulling in and bolting the edge ropes, the arches can be pulled apart using rigging ropes or webbing slings with tirfors and fixed (2 in Fig. 348). The surface can be finely tensioned by shortening the edge rope (3 in Fig. 348). To tension a number of arch fields, each field is tensioned alone and secured with temporary construction.

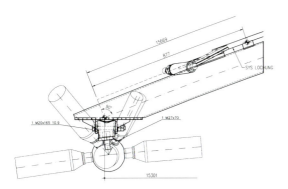 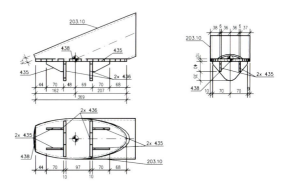

Fig. 349: Arch bearing at the stadium in Al Ain, Dubai, United Arab Emirates

Fig. 350: Erection of the stand roofing at the stadium in Al Ain, Dubai, United Arab Emirates

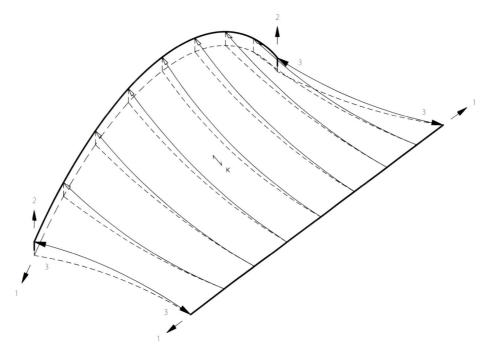

Fig. 351: Pretensioning a fabric membrane by vertical movement of the arches

Introduction of loads by moving the edge beams

Linear displacements in the membrane surface can also be achieved with arched structures by pushing up the edge arches.

This is done by first prestretching the membrane edge in the direction of the arch and then clamping it to the arch (1 in Fig. 351). After pulling in and bolting the edge rope, the arches can be jacked up in stages by hydraulic presses, which pretensions the membrane (2 in Fig. 351). The edges can subsequently be finely tensioned by shortening the edge ropes (3 in Fig. 351).

Whether this method can be used depends on the force necessary to introduce the load in relationship to the rise height, the form of the valley and the cutting pattern direction.

If high forces have to be used to push up the arches on account of the strongly shortened membrane surfaces, then this method is uneconomical.[1]

If a row of arch fields are arranged parallel to each other, the membrane is mostly laid over the arches without fixing to the crown of the arch.

Fig. 352: Erection of the stand roofing at the Fenerbahce stadium in Istanbul, Turkey

1 Lenk, S. (2004)

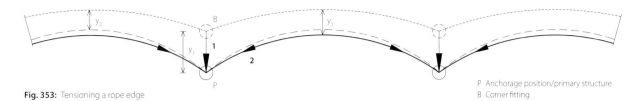

Fig. 353: Tensioning a rope edge

P Anchorage position/primary structure
B Corner fitting

3.5.3.2 Application of loads at points

Flexible edge details with ropes or belts running in sleeves can be pretensioned by displacing the stiff corner fittings (B in Fig. 353) at points. This can be direct pulling or pushing of the fitting to the anchorage position. With awnings, the displacement is mostly produced by shortening the stay ropes, which causes a rotation of the columns.

Tensioning by direct pulling or pushing is performed in many stages. After the edge rope has been pulled in and the corner fittings preassembled, these are displaced towards the fixed primary construction element (1 in Fig. 353). After an alteration of position by distance y_1 or y_2 and with the reaching of the anchorage position (P in Fig. 353), the corner fittings can be hung from the primary structure. In the second stage, the edge ropes are finely tensioned to the corner (2 in Fig. 353). The rope edge has then moved a distance y_3 at its crown.

Corner constructions with elements like tensioning bolts or rollers can make the tensioning process easier. Heavy constructions are mostly mounted on movable supports (right in Fig. 355).

In order to apply the load at a point into membrane surfaces, the geometry of the corner is important. The more acute the angle is at the membrane plate, the less is the available material capacity (see section 2.7).

High stresses in the fabric cannot be relaxed there through stretch and angular rotation on account of the short distance to the edge. Insensitive tensioning or imprecise patterning in the area of the membrane plate can therefore lead to extraordinary peaks of stress in acute corners under tensioning. The danger of the fabric tearing at the corner is mostly due to such peaks and not the size of the membrane fields to be tensioned. If the breaking tension is exceeded during tensioning, this would cause considerable damage to the fabric. It can be advisable to increase the material capacity through fabric reinforcements.

Fig. 354: Tensioning a rope edge

Fig. 355: Pretensioning a fabric membrane by displacing the corner fitting

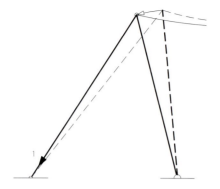 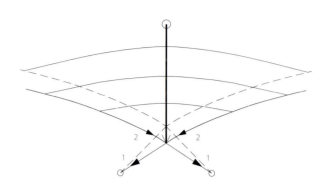

Fig. 356: Tilting of masts by shortening the stay ropes

Application of load by tilting of shaft-shaped supporting elements

Point loads can also be introduced into membrane surfaces by displacing or rotating the stiff primary elements. This tensioning process is frequently used for the erection of awnings, where the fabric is pretensioned by the tilting of the masts and the shortening of the stay and edge ropes (Fig. 356).

The fabric is fixed to the mast head by connecting the fittings and hanging the edge rope. In order to introduce the main pretension, the mast can be tilted through the use of tensioning devices with hydraulic presses or using ratchet pullers, depending on the level of the forces to be introduced (Fig. 357). Care must be taken that the tilting is exactly in a plane, so that the eye lugs at the foot of the mast suffer no deformation. The masts need to be stabilised sideways during tilting.

The tensioning is done in the sequence laid down in the erection planning and depends on local conditions, cutting pattern direction and fabric material. The duration of tensioning and the sequence are especially critical when tensioning glass fibre fabrics, where delayed tensioning can enable the stress to distribute in the surface. The force level can be balanced by alternatively tensioning and relaxing.

Because of the large deformations, the rope ends hung at the mast head represent a particularly sensitive area when tilting the masts. It is important that the ropes do not suffer kinking where they emerge from the fittings.

The arrangement of drillings in the foundation linkages for the stay ropes can make the tilting operation much easier (left in Fig. 357). Connecting members and temporary stays can also be fixed to these.

To finely adjust the masts, the stay ropes are shortened with turnbuckles. The fine tensioning of the edges is done by shortening the edge ropes.

Fig. 357: Erection of the roofing of the square in Zeltweg, Austria

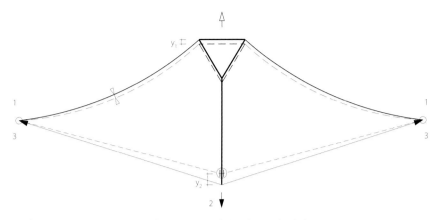

Fig. 358: Central tensioning of a high point supported by ropes from underneath by lengthening the shaft axis

Application of load by enlargement of shaft-shaped supporting elements

Structural elements in steel construction can be pretensioned by processes, which lengthen or shorten the effective axial length of structural elements through the installation of mechanisms in the construction.[1] This method can also be used to pretension very efficiently in membrane construction. High point constructions with standing or hanging columns can be pretensioned centrally by vertically jacking the columns and introducing forces into the membrane surface through the connected edges. Fig. 358 shows the displacements at column end points y1 and y2 during this tensioning process. They result from the geometrical stiffness, the curvature, the differing strain stiffnesses of ropes and fabric and the level of the forces introduced.

To be able to lengthen the effective axial length, the column has to be made in more than one part. The column can either be connected through an intermediate member with an adjustable column foot or pushed through the column foot. After fixing the membrane to its edges at corners (1 in Fig. 358), the column is pushed apart in stages by hydraulic presses and the final length fixed with bolts (2 in Fig. 358). If no presses are available, columns hanging from a crane can be ballasted at their foot point. The position of the column can be fixed by retensioning the rope supporting it from underneath (3 in Fig. 358).

One disadvantage of this tensioning method is that the entire pretension force is applied at one point. This means that relatively high forces have to be used. Another way of pretensioning high point structures is to tension the membrane solely by peripheral pulling of the edges and corners (b, d in Fig. 359). This requires less force to be applied and a better stress distribution in the surface can be expected, but the amount of work involved in installing equipment for stretching along the edges and at the corners can cause high costs.

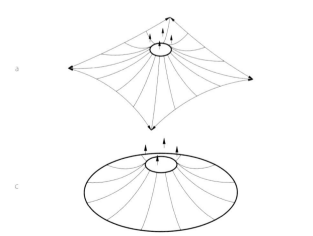
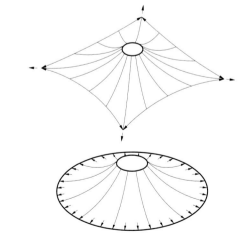

Fig. 359: Central and peripheral tensioning of high point constructions

---

1 Ferjencik, P.; Tochacek, M. (1975)

In order to avoid stress concentrations at the high point, the high point radius should be designed to be large for both tensioning methods. Otherwise it would be necessary to reinforce the fabric in this area.

An example for the central introduction of pretensioning forces in a high point structure supported from underneath by wires is the roofing over the vehicle park at the Munich Waste Management Office (left, right in Fig. 360). The Glass/PTFE fabric for the roofing was prefabricated at the works in 10–12 m widths and 70 m long strips with seams in the surface of 60 mm. The strips were laid out in the transverse direction on site and welded to the adjacent strips (see section 2.5.1.1). In order to be able to compensate for errors, the weld seams were made on site with a width of 150 mm. During the welding work, the hats of the high points were turned inwards and fixed down against wind loads, and in order to avoid puddle formation through rain or snow (right in Fig. 361).

Then the inverted hats were turned out upwards and the hanging columns were fitted from underneath (this method of installation incidentally only works with appropriately large radius and spacing of the high points). After lifting the hanging columns (right in Fig. 361), the ropes to support them from underneath were pulled in and anchored to the tree column. The staged pushing apart of the hanging columns by up to 300 mm introduced the pretension into the fabric and ropes. After adjusting the ropes, the columns were fixed in their final positions. For the duration of the erection process, the structure was stabilised by tirfors and ratchet tie-downs, which were remove after completion of tensioning.

An endless rope was fabricated into the membrane at the upper ring and secured with 3 pegs. This construction ensures that when sideways movement occurs, no sharp edges can damage the fabric and the rope can roll of the ring under loading. There are no clamping plate connections between the 10 x 12 m panels.

Fig. 360: Erection of the compartmental high point surface at the Munich Waste Management Office, Germany

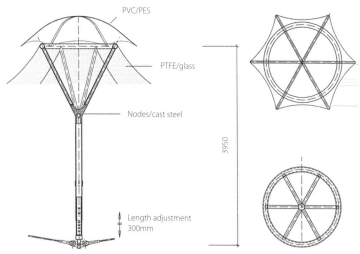

Fig. 361: Installation of hanging column at the Munich Waste Management Office, Germany

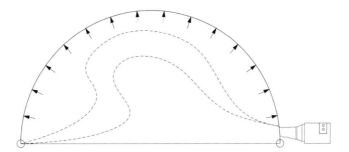

Fig. 362: Pretensioning by creating pressure differences

3.5.3.3 Application of surface loads

Synclastic surfaces cannot be pretensioned by tangential displacement of elements in the surface. They have to be tensioned by applying surface loading perpendicular to the plane. The pretensioning of rope hanging roofs is done by ballasting. This creates a loading condition, which ensures that no compression will occur in the tension elements through external influences like wind uplift or snow. The application of surface loading by ballasting can be seen not as actual pretensioning, but as a loading condition.

Pneumatic construction makes use of the medium air as pressure dissipating element for the envelope. The surface loading results from the low difference of pressure between the air inside and outside. The forms of pneumatic constructions always follow the general formula for the stress in the walls of a pressure vessel. Loads acting inwards are supported by the pneumatic envelope and lead to a reduction of the stress in the membrane. Loads acting outwards are distributed in the plane of the envelope and mostly increase the stress.[1] The types of pneumatic construction are air-supported halls, large pillows and compartmental pillows. Suitable materials for the envelope are PVC-coated polyester fabrics and, for limited spans, ETFE foils.

Air supported halls have to be anchored around the perimeter foundation. The form of the foundation line has to correspond to the correct form of the edge of the envelope under long-term stress. The factors, which determine the edge detail, are the stretch properties of the envelope material and the surface curvature in the edge area.

The envelope expands in all directions during the inflation process. The enlargement is, however, not uniform. It can happen that the stretch in one direction is very large while the material in the other direction tends to shrink.[2] The risk of wrinkles forming in the area near the edge therefore has to be investigated. A high arch rise and steeply inclined side walls lead to strong deformation under loading, which load the anchoring elements at the edge. The resulting rotations at the bearing have to be taken by the edge element. The standard edge detail for air-supported halls is a steel tube in a membrane sleeve. Other possible anchorages are clamping plate edges and laced edges at the foundation.

Flat dome forms can be connected to a bearing compression ring, which can also take horizontal forces. Ballast bodies for temporary foundations can be fabric tubes filled with earth, sand, gravel or excavation spoil, or precast elements filled with water (Fig. 363).

To maintain the air pressure in pneumatic structures, blower units are used, which can be regulated and controlled. These supply the membrane envelope though a duct system with

Fig. 363: Anchoring to water containers

---

1 Stimpfle, B. (2004-1)   2 Otto, F.; Happold, E.; Bubner, E. (1982)

Fig. 364: *left:* Inflation connections for an air-supported hall; *right:* Blower plant for a large pillow

an excess pressure of 200 – approx. 500 Pa. One exception is high-pressure tyres, which are supplied with pressures of about 0.5 bar.

Several blowers are normally used together to supply air-supported halls and pillow constructions, blowing air into the hall either alternatively or through a common connection. When they work alternatively, non-return flaps are installed in the blower ductwork, so that if one blower stops the air will not escape back though it. In order to protect the material of the envelope, higher pressures should only be used when the corresponding loading case arises. When the pressure falls under the required minimum value, further blowers can be switched on to give support. A pressure transmitter can be installed for pressure measurement.

In order to achieve the best possible transparency, ETFE foils are being increasingly used for constructing pillows. Inflow of moisture can, however, obscure the light coming through. In order to make sure that as little moisture gets into the pillow as possible, there should be a dehumidifier before the blower. The inclusion of a condensate separator, which has to be checked and emptied regularly, can prevent rainwater or condensate forming in the pressure measurement hoses.

The erection of air-supported halls starts with the construction of the foundations and the assembly of the inflation equipment. After the protected laying out of the membrane envelope, the assembly joints are made and the membrane is clamped to the edge of the lock and the rings of the blower connections. Once the perimeter edging has been fixed and any ballast bodies have been filled, the preparation works for inflation are complete. The pressure monitoring system is activated and the inflation can begin.

During inflation, the pressure must be controlled constantly using a pressure display. The inflation process for halls of average size is normally finished in less than 1 hour. The loading of the connections should be watched. Inflation cannot be done with a strong wind. Kinking of the envelope during inflation must be avoided. If it rains or snow falls during inflation, there is a risk that puddles could form. If the material of the envelope stretches differently when inflated, then biaxial tests should be performed, and the resulting measures for inflation are important.

The roof should periodically be inspected and checked in an optical control of the roof construction. The membrane should be checked for air-tightness, air pressure and material condition. The inlet filter in the switching cabinet and the drier filter should be cleaned or replaced. The pressure measurement hoses should be checked for condensate formation and blockage.[1]

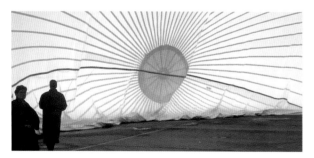 

Fig. 365: Fabric membrane during the inflation of an air-supported hall

1 *Nolting-company material (1999)*

3.6 Control of the forces in flexible structural elements

The assembling of different materials and the involvement of all parts of the structure in load transfer places high demands on the geometry of lightweight structures. The deformations of the load-bearing elements in tension resulting from form-active interaction limit each other mutually. The coordination of the differing strains of the materials is therefore fundamental for the load-bearing behaviour of structures loaded in tension. This situation demands that measurements of force and geometry are performed on the individual parts of the structure during the tensioning process. In the course of applying the loading, unintended load transfer could lead to constraint forces and can lead to local overloading and destruction of the ropes. In order to avoid this, force measurement members are provided between the primary construction, tensioning tools and the flexible load-bearing elements, and the values occurring during tensioning are compared with the intended values and recorded. The final state and the geometry also have to be monitored. In cases where long-term monitoring is needed, permanently installed measurement devices are used.

There are a number of processes for the measurement of rope forces, but it is not always possible to determine an ideal method of measurement. The suitability of a measurement process depends mainly on the rope measurements, the degree of pretension, the work involved in installation and local conditions for force measurements. Often several methods are combined for one structure in order to achieve economically justifiable and sufficiently exact measurement results. For example, four different measuring procedures were used for the measurements of rope forces during the erection of the Forum roof at the Sony Center in Berlin.[1]

In addition to the type of rope, the rope diameter and the expected force in the rope, other criteria like current temperature, wind conditions and manageability can influence the selection of measuring method.

Forces in structural elements are not directly measurable, they can only be defined by their effect on suitable measuring equipment. Because of the dependence of the force on the mass and acceleration of a body, there is a direct relationship of force to defined physical values like elasticity (strain, extension), pressure or piezo-electricity.

A commonly used method for measuring force is the exploitation of the elastic deformation of solid bodies. Mechanical, hydraulic or electrical measuring devices record the force to be measured through deformation of the measuring element directly in the flow of force. The force measurement can be undertaken with measuring elements, which work with elongation, bending or shear.[2]

Further methods of checking rope forces are the setting off of vibrations in the rope, where the force is calculated from the measured resonance frequency, and the measurement of the static deflection, from which the force can be calculated if the rope length, cross-sectional diameter and material data of the rope are known.

The best-known process in strain measurement technology is measuring with strain gauges. These tolerate a high number of load cycles, only have a small dead weight and are used in a multitude of applications as measurement sensor. They are applied by sticking them to the body to be measured.

Strain gauges work on the basis of the alteration of the resistance of an elongated wire or strip of metal foil. The meander-shaped resistance wire, which is fixed to a carrier foil or a corresponding metal foil, which is similarly structured by etching or vaporisation, is elongated by the strain. This makes the cross-sectional area of the wire smaller and alters its crystalline structure. This effect on the wire alters its electrical resistance and this can be calculated from the specific resistance of the material and its length.

The disadvantage of the strain gauge for measuring the strain in ropes is lack of practicality, considering the often uneven surface and the measurement results. It is often unclear which strain is being measured, because the surface of a rope can show a different strain then the strands or the insert. For measuring the strain in ropes, the strain gauge is mostly installed in force measurement boxes.

3.6.1 Determination of force in ropes

One method of determining the forces in ropes is the measurement of strain, with the rope itself being used as force measurement element. To do this, an extensometer is attached to the rope in the unstressed state, which measures the initial length and the length under tension. This enables the force in the rope to be calculated from the strain and the strain stiffness. A whole range of piezoelectric or inductive sensors or also vibration sensors can be used for this process. It is essential that the sensor is clamped to the rope in a stress-free state and is not disturbed during erection (a in Fig. 366).[3]

---

[1] Lindner, J.; Breitschaft, G.; Thaten, J. (2000)

[2] Labor Blum company material

[3] Labor Blum company material

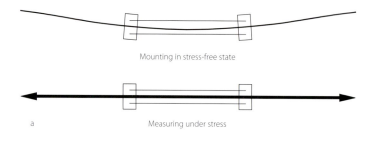

Fig. 366: Measurement methods for rope force determination through measurement of strain

a — Measuring under stress
b — Relief method

Another way of determining the force in a rope is to mount the extensometer on the pretensioned rope and then to unload the rope (b in Fig. 366). The force is determined analogously to the above example. This method can, however, only be used with suitable tensioning apparatus and is therefore restricted to relatively thin ropes and rope forces.

Edge ropes in rope nets mostly have large diameter and high rope forces, which make an unloading unnecessary. In this case it is possible to measure the forces in the net rope running into the edge rope using the unloading method and determine the force out of the equilibrium conditions with the geometrical and loading values for net and edge rope.[1]

Direct or indirect methods of measurement can be used to determine the tension forces in ropes. A direct method of measuring the tension forces can be carried out with dynamometers installed between hydraulic presses or by measuring the pressure with a manometer. Indirect methods can mean measuring the geometrical deflection sag or by measuring the resonance frequency.[2]

Measuring with hydraulic force transducers

The best-known method of determining the force in ropes in steel erection is the measurement of the compression in hydraulic cylinders installed in the flow of the forces during the pretensioning process. Various intermediate measurements are undertaken in the course of force introduction into the rope by hydraulic presses controlled according to force or travel. The compression force in the press is read from a calibrated manometer until the permissible press lift as specified in the erection plan has been reached, and is recorded together with the relevant piston travel. The applied force can be calculated with knowledge of the data for the device (area of the press cylinder). After the full pretension has been reached and the press compression has dropped off, the tensioning equipment can be removed. It is often necessary to readjust the rope length when using this process.

One disadvantage of this method is the size and weight of the presses and the associated work in setting up in a new location. The high weight of presses can well require use of a crane.

Vibrating wire process

In this electrical measurement process, the resonance frequency of a string clamped into a holder is measured during alteration of length and converted. When a tension force is applied, the tension in the wire changes and thus its resonance frequency. The wire can be resonated by a piezoelectric or an electromagnetic component transverse to the

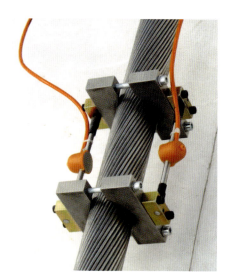

Fig. 367: Vibrating wire sensor

---

1 Labor Blum company material

2 Ramberger, G. (1978)

longitudinal axis, the measured electrical voltage converted and fed to an electronic counter device.

The vibrating wire sensor is clamped to the rope in a stress-free state. Stress-free ropes have a slight bend. In order to compensate for this bending and achieve exact values, it is necessary for this process to mount 2 sensors in one plane (Fig. 367). The strain in the rope then results from the average resonance frequency value of the shortened and the lengthened wires. The advantages of this method are long-term stability and high precision. It is therefore often used for long-term monitoring. One disadvantage is that the sensor needs a relatively high setting force for adjustment.[1]

Three-point bending measurement

This method of measurement is performed by applying a deflection of defined extent to the rope and measuring the required force. The rope is laid on saddles and is given a defined angle by a hand-operated central tensioning shoe. If the rope is under load, the force in it works against the angle of deviation and acts upon a built-in electric load cell, which sends a signal dependant on the force to a process-controlled digital display (Fig. 368). This measurement is not suitable for long-term use.

For the measurements in connection with the rope façade of the Cologne airport, a rope force meter based on the three-point bending test was further developed, which in addition to the force can also measure the deflection. This makes it possible to work with smaller deflections. Development work is also aimed at using hinged saddles to distribute the unavoidable increase of strain at this location more uniformly over the entire length of the rope. A built-in microprocessor digitises the measured data of force and deflection and makes it available for further processing. This device can be used to measure ropes up to 38 mm diameter and forces of 250 kN.[2]

3.6.2 Measurement of membrane stresses

In order to avoid permissible stresses being exceeded, check measurements have to be performed on installed and tensioned membranes. The measurement of membrane stress is, however, problematic.

Measurement of the strain with strain gauges is difficult to handle and evaluate. The strip has to be attached to the membrane surface when this is stress-free in order to be able to determine the strains after the loading has been applied. It is hardly practical to protect the sensitive strain gauge during the process of erection. The evaluation of the strains in order to determine the stress is also very difficult on account of the anisotropic and viscoelestic behaviour of the fabric. Other procedures for measuring membrane stress also have problems with the curvature of the membrane surface and the inhomogeneous (locally not constant) stress conditions to be expected.[3]

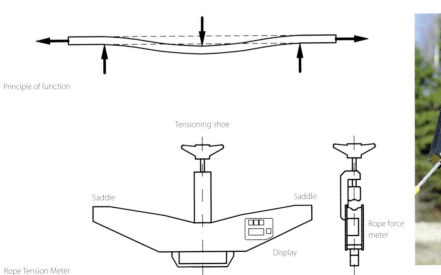

Fig. 368: Three-point bending measurement

1 Labor Blum company material

2 Dürr, H. (2000)

3 Blum, R. (1982)

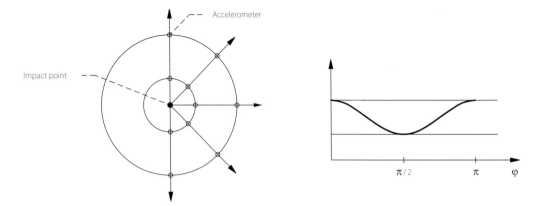

Fig. 369: *left:* Arrangement of the accelerometer; *right:* Curve relationship with ray angle

Three methods of measuring stress developed by Labor Blum are described below

The first method is based on the observation of the spread of acceleration waves, which start from a short impact on the membrane surface. Two concentric circles are drawn round the impact point at defined diameters and accelerometers are fixed at the intersection of the circles with rays projecting from the centre (Fig. 369). After the impulse has been generated, the movement condition of the membrane is recorded, forwarded to an oscilloscope as electrical signal and displayed on a screen. On the basis of three measurements in any direction, the maximum and the minimum time and thus the speed of the elliptical spread of the wave, which is needed as the basis for the calculation of stress.[1]

In the second process for determining the stress situation, a pressure depression is made in the surface by applying a defined force. A plate-shaped device is used for this, which presses a plunger within a ring against the membrane surface (Fig. 370). The deflections and stresses arising in the membrane surface can then be calculated.

The third measurement method describes a measuring device developed for test purposes at Labor Blum, which is constructed with two potentiometers, a ring force measuring device, a turning disc and an annulus with press-down ring.

To perform a measurement, the base ring is pressed onto the membrane by a partial vacuum and a defined point load is applied in the axis of the base ring. The rotating disc with two potentiometers mounted on it is set turning and the deflection of the concentric rings is measured. During the rotation, the two potentiometers show the deflection curve as a function of radius, with which the relationship to the defined load can be calculated (Fig. 371).[2]

Fig. 370: Plate method

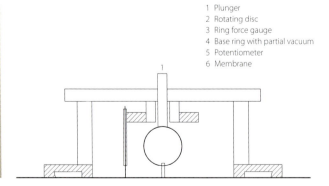

1 Plunger
2 Rotating disc
3 Ring force gauge
4 Base ring with partial vacuum
5 Potentiometer
6 Membrane

Fig. 371: Ring force measuring devicer

1 Blum, R. (1982) *2 Labor Blum company material*

4 Summary and outlook

The present book has attempted to give an overview of the current state of the art in the implementation of tensile surface structures. The purpose of the work has been to permit the recognition of fundamental interactions between the production and the erection of flexible structural elements.

The investigation of the factors affecting the construction process, as result from the manufacture of flexible structural elements, illustrates the interdependency of production and erection processes. The emphasis has been the production and assembly technology of coated textile sheet materials.

These investigations have been based on the collection and structuring of the highly specialised knowledge of designers, manufacturers and erectors. This, and also the classification and definitions of the criteria relevant for construction, can be seen as the main contribution of this book to the practical implementation of wide-span, lightweight structures

Starting from the textile characteristics and internal structure of flexible materials, the first main part of the book describes their composition and manufacture and explains how the technicalities of their manufacture determine the complex material behaviour and how they function as structural elements A summarised description of the mechanical behaviour of coated fabrics was necessary for the evaluation of the influence of the materials used as flexible structural elements on the construction process. In addition to the internal structure of coated fabric, which is constructed to meet its specific purpose, it is mainly the weave type used in the production process, which has a substantial effect on the deformation behaviour under the influence of load, time and temperature during erection. The interaction of the thread directions depending on the calculated compensation value is of central importance for the erection of membrane surfaces.

Another example of the reciprocal influence of the mechanical properties of the material between production and erection is the patterning of textile sheet elements. The definition of criteria for the patterning showed how the determination how the design of the spatial cutting-out pattern of the individual fabric strips on the membrane surface affects the load transfer and the erection. It was demonstrated here what effects the selection of material and also the determination of the tensioning direction and the tensioning sequence have on the erection of membrane surfaces. It can be stated regarding the topographical and structural criteria that the type of material, the shape of the surface, strip layout, strip shape, seam geometry, seam arrangement and the geometry and type of edging all interact, which considerably affects the behaviour and interaction of the structural elements.

Statements about these relationships are, however, only possible after detailed investigations and tests. Regarding the parameters of form and load-bearing behaviour, which affect the patterning, there is a need for more detailed analytic investigations into this complex area of work. One possible starting point would be to research the relationship of rise height and main curvature to possibly achievable spans and strip lengths of anisotropic membrane surfaces.

A further section illustrates in the first part examples of ways of jointing the surface and methods of introducing force into the edge, and their effects on the structural behaviour and special factors affecting manageability and ease of erection. The starting point for further considerations in this area was the detailing of corner, edge and surface connections so as to make them buildable and manageable.

The second main part dealt with the erection of flexible structural elements. The description of the economic and technical conditions for construction management looked into the planning forecast of the construction process. The factors resulting from production practicalities on the scheduling and the influences and areas of responsibility of construction planning concerning the design and construction of membrane structures. Ways of modelling the erection activities and aspects of detailing in such a way as to increase the practicality of erection were also discussed.

An illustration and a description of the equipment used describes the role of the tools and equipment used in transport, lifting and tensioning, and explains the resulting parameters for the design process.

Building on the foundation of production and construction technology parameters, the erection principles and influential factors affecting the erection process were defined and investigated. In addition to the type and the jointing of the materials used, the local conditions on the construction site, the type of equipment used for erection and the practicalities of production, delivery and transport, it is above all the working principle of the structural system, which decisively influences the choice of a suitable erection procedure. The emphasis of this section is therefore a description of temporary states during the erection of form-active structural systems seen from the practical point of view. The nature of the interaction of primary structure, sub-structure and secondary structure elements was investigated regarding the erection technology. Stabilisation measures for the erection of form-active structures were discussed through a comparison of the loading of structures and structural elements during erection with parameters from the safety and compensation plan, taking construction technology into account.

In a further section, erection procedures for mechanically tensioned membrane structures were described and illustrated with examples, and the erection sequence and construction activities for the erection of characteristic forms of structure were explained. In order to do this, it was also necessary to go into construction problems in the areas of delivery, handling and jointing and categorise and describe the individual erection implementation activities. The documentation of completed projects and schematic diagrams visualises these themes.

The main emphasis of the discussion of the construction process was the categorisation and description of the procedures for introducing the pretensioning into flexible structural elements from the practical point of view. The tensioning of flexible structural elements is of central importance in the erection of membrane structures. Sufficient stability of the structural element and the structural element is not created until the tension has been applied to the linear and sheet elements. The growth and creation of the structure through the successive introduction and constant increase of the pretensioning forces differentiates the erection of wide-span, lightweight structures intrinsically from conventional construction procedures, where the erection of the structure occurs through constant addition of new elements. In addition to the preassembly, lifting and assembly, the tensioning process is a major characteristic of the erection of wide-span lightweight structures.

Finally, an overview was presented of processes for measuring forces in ropes and membranes on the construction site.

In summary, it can be observed that an overall assessment of the erection of wide-span lightweight structures is very complex, on account of the multitude of influential factors. This type of structure is a unique product, with each new project producing new conditions and different aims for erection. The main parameters affecting the erection of membrane structures are the deformation behaviour of the material used, its jointing and edging and the determination of the spatial cutting patterns. The manner of interaction with the primary structure is also essential for the installation of the flexible elements.

In the opinion of the author, great potential for optimisation of the erection lies in the production of the sheet textile elements. New developments in materials, which offer more satisfactory solutions for the elastic and load-bearing behaviour under the effects of load, time and temperature, jointing methods and the requirements arising from the load-independent influences on the material, could considerably simplify the construction technology. Above all the deformation behaviour of coated fabrics and the effect of it on the tensioning process could offer potential for future development.

In summary, it can be observed that the erection of form-active structures is a very complex field. In order to avoid mistakes and shortcomings in the implementation on site, the importance of erection practicalities should be awarded sufficient importance at an early stage by architects and engineers responsible for the design of wide-span lightweight structures

References

Albrecht, R. (1973) | Montagelehre (Construction practice), Ernst & Sohn, Berlin

Alpermann, H.; Gengnagel, C. (2003) | Interaktion von Membran und biegesteifen Bogentragwerken (Interaction of membrane and stiff arch structures); in: Stahlbau 72, Heft 10, Ernst & Sohn, Berlin, p. 702 – 707

Baumann, Th. (2002) | Statisch-konstruktive Eigenschaften von PTFE-beschichtetem Glasgewebe (Structural and constructional properties of PTFE-coated glass fabrics); in: Bauingenieur 77, Springer-Verlag, p. 158 – 166

Baumüller, D. (2000) | Rotationspneu (Rotating pneu roof); in: Detail 6/2000, Institut für internationale Architektur-Dokumentation, München, p. 1086

Beck, W. (1990) | Seilverguß (Rope speltering); in: Feyrer, K. u. a., Stehende Drahtseile und Seilendverbindungen, Expert-Verlag, Ehningen, p. 103 – 119

Bergermann, R.; Sobek, W. (1992) | Die Roofing der antiken Arena in Nîmes (The roofing of the ancient arena in Nîmes); in: Bauingenieur 67, Springer-Verlag, p. 213 – 220

Bergermann, R.; Göppert, K.; Schlaich, J. (1995) | Die Membranüberdachungen für das Gottlieb-Daimler-Stadion, Stuttgart, und den Gerry-Weber-Centre Court, Halle (Westfalen) (The membrane roofing of the Gottlieb Daimler Stadion, Stuttgart, and the Gerry Weber Centre Court, Halle, Westphalia); in: Bauingenieur 70, Springer-Verlag, p. 251 – 260

Blum, R. (1982) | Spannungsmessungen in vorgespannten Membranen (Measurement of stress in pretensioned membranes); in: Otto, F.; Happold, E.; Bubner, E. (Hrsg.), Lufthallenhandbuch, Karl Krämer Verlag, Stuttgart, p. 268 – 274

Blum, R. (1990) | Zeltbaumaterialien (Materials for tent construction); in: Günther Brinkmann (Hrsg.), Leicht und Weit. Zur Konstruktion weitgespannter Flächentragwerke, Ergebnisse aus dem Sonderforschungsbereich 64, VCH-Verlagsgesellschaft, Weinheim, p. 200 – 224

Blum, R. (2002-1) | Material Properties of Coated Fabrics for Textile Architecture; in: Marijke Mollaert (Hrsg.), The Design of membrane and lightweight structures, VUB-Press, Brüssel, p. 63 – 88

Blum, R. (2002-2) | Bericht über die Denkmaleinhausungen in Clemenswerth und Weikersheim. Projekt "Winterzelt" (Report of the housing for the ancient monuments in Clemenswerth und Weikersheim. Project "Wintertime"); aus: http://www.echn.net/conservation/epuclications

Blum, R. (2004) | Nicht veröffentlichtes, transkribiertes Tonbandprotokoll eines Interviews im Labor Blum in Stuttgart (Unpublished transcription of a taped interview in the Blum laboratory in Stuttgart)

Blümel, S.; Stimpfle, B.; Rudorf-Witrin, W.; Pasternak, H. (2005) | Von der CargoLifter-Werfthalle zu Tropical Islands – Entwurf (From the CargoLifter works hall to Tropical Islands – draft); in: Bauingenieur, Band 80, Oktober 2005, Springer-VDI-Verlag, p. 461 – 469

Bögner, H. (2004) | Vorgespannte Konstruktionen aus beschichteten Geweben und die Rolle des Schubverhaltens bei der Bildung von zweifach gekrümmten Flächen aus ebenen Streifen (Pretensioned construction and the role of shear behaviour in the formation of doubly curved surfaces out of flat strips); Dissertation an der Universität Stuttgart, Fakultät Bau- und Umweltingenieurwissenschaften, Institut für Werkstoffe im Bauwesen

Böhmer, C. (2004) | Nicht veröffentlichtes, transkribiertes Tonbandprotokoll eines Interviews bei der Fa. Montageservice-SL GmbH in Hallbergmoos (Unpublished transcription of a taped interview at the company Montageservice-SL GmbH in Hallbergmoos)

Bubner, E. (1997) | Membrankonstruktionen – Verbindungstechniken (Membrane construction – jointing techniques), Verlag Wehlmann, Essen

Buckminster Fuller, R. (1973) | Aufsatz: Einflüsse auf meine Arbeit; in: Krausse, J. (Hrsg.), Bedienungsanleitung für das Raumschiff Erde und andere Schriften (Essay: Operational instructions for the spaceship Earth and other writings), Rowohlt Verlag, Reinbeck

Bukor, S. (2003) | EXPO SUISSE 2002, Studienarbeit am Institut für Architektur und Entwerfen, Fachbereich Hochbau 2

Burkhard, W. (1998) | Textile Fertigungsverfahren (Textile manufacturing processes), Carl Hanser Verlag, München

Cenotec (1999) | http://www.ceno-tec.de/ind05frd.htm, Greven

DIN 536-1 bis DIN 15030 (1995) | Krane und Hebezeuge 1 (Cranes and lifting devices 1); DIN-Taschenbuch, Deutsches Institut für Normung, Beuth, Berlin, Vienna, Zürich

Domininghaus, H. (1992) | Die Kunststoffe und ihre Eigenschaften (Plastics and their properties), VDI-Verlag, Düsseldorf

Drees, G.; Krauß, S. (2002) | Baumaschinen und Bauverfahren, Einsatzgebiete und Einsatzplanung (Construction machines and processes, scope of use and planning of use), Expert Verlag, Renningen

Dürr, H. (1988) | Montagekonzept zur Roofing der antiken Arena in Nîmes, nicht veröffentlicht (Erection concept for the roofing of the ancient arena in Nîmes, unpublished)

Dürr, H. (2000) | Seilnetze-Planung, Berechnung, Ausführung und Detailed design (Rope nets, calculation, implementation and works design), in: Stahlbau 69, Heft 8, Ernst & Sohn, Berlin, p. 585–594

Dürr, H. (2002) | Montagepläne Galet 1 – Expositon Nationale Suisse 2002, Arteplage de Neuchâtel (Construction plans for Galet 1, Swiss national exhibition at the Arteplage de Neuchâtel)

Dürr, H. (2003) | Nicht veröffentlichtes, transkribiertes Tonbandprotokoll eines Interviews bei der Fa. IF-Group in Reichenau (Unpublished transcription of a taped interview at the company IF-Group in Reichenau)

DVS-Richtlinie 2225 (1991) | Fügen von Dichtungsbahnen aus polymeren Werkstoffen – Schweißen, Kleben, Vulkanisieren (Jointing of waterproofing material from the roll – welding, glueing, vulcanising); in: Fügen von Kunststoffen, DVS-Merkblätter und -Richtlinien, Fachbuchreihe Schweißtechnik, DVS-Verlag, Düsseldorf

Eberspächer-Firmenschrift (2003) | Hochdruck-Hydraulik-Datenblätter (High-pressure hydraulics data sheets), Kirchheim/Teck

Engel, H. (1997) | Tragsysteme, (Structure Systems), Hatje Cantz, Stuttgart

Essrich, R. (2004) | Nicht veröffentlichtes, transkribiertes Tonbandprotokoll eines Interviews bei der Fa. IF-Group in Reichenau (Unpublished transcription of a taped interview at the company IF-Group in Reichenau)

Ferguson, E.S. (1993) | Das innere Auge. Von der Kunst des Ingenieurs (The inner eye, the art of the engineer), Birkhäuser Verlag, Basel, Boston, Berlin

Ferjencik, P.; Tochacek, M. (1975) | Die Vorspannung im Stahlbau, Theorie und Konstruktionspraxis (Pretensioning in steelwork, theory and construction practice), Bauingenieur-Praxis, Heft 38, Verlag Wilhelm Ernst & Sohn, Berlin

Feyrer, K. (1986) | Spleißen; in: Spur, G.; Stöferle, Th., Handbuch der Fertigungstechnik; Band 5 – Fügen, Handhaben, Montieren (The handbook of manufacturing technology; volume 5 – jointing, handling, mounting), Carl Hanser Verlag, München Vienna, p. 99–104

Fitz, H. (1989) | Witterungsbeständige Fluorkunstoffe im Außeneinsatz (Weatherproof fluoroplastics in outdoor use); in: Werkstofftechnik, Carl Hanser Verlag, München, p. 519–524

Fitz, H. (2004) | Nicht veröffentlichte Korrespondenz (Unpublished correspondence)

Föll, H. (2005) | Elastische und viskoelastische Eigenschaften von Polymeren (Elastic and viscoelastic properties of polymers), Vorlesungsskript des Lehrstuhls für allgemeine Materialwissenschaft an der Universität Kiel

Fritz, C. P. (1999) | Struktur und Eigenschaften von Polyester-Mischungen und daraus erstellten Fasern (Structure and properties of polyester mixtures and the resulting fibres); Dissertation an der Universität Stuttgart, Fakultät Chemie, Institut für Chemiefasern der Deutschen Institute für Textil- und Faserforschung, Stuttgart

Funk, J. (2005) | Nicht veröffentlichte Korrespondenz mit der Fa. P-D Interglas-Technologies AG (Unpublished transcription of a taped interview at the company P-D Interglas-Technologies AG)

Gabriel, K. (1990) | Konstruktion und Bemessung (Design and detailing); in: Feyrer, K. u. a., Stehende Drahtseile und Seilendverbindungen, Expert-Verlag, Ehningen, S. 1–42

Gabriel, K.; Wagner R. (1992) | Bauen mit Seilen (Building with ropes), Vorlesung am Institut für Tragwerksentwurf und Konstruktion, Wintersemester 1991/92, Universität Stuttgart

Göppert, K. (2003) | Membrankonstruktionen – Haut und Knochen (Membrane construction – skin and bone); Vortrag bei der Internationalen Vortragsreihe am Institut für Hochbau 2, TU Vienna

Graf, W. (2004) | Nicht veröffentlichtes, transkribiertes Tonbandprotokoll eines Interviews bei der Fa. Sattler AG, Graz (Unpublished transcription of a taped interview at the company Sattler AG, Graz)

Gropper, H.; Sobek, W. (1985) | Zur konstrutiven Durchbildung ausschließlich zugbeanspruchter Membranränder (About the construction details of membrane edges loaded purely in tension) ; in: 3. Internationales Symposium des Sonderforschungsbereiches "Weitgespannte Flächentragwerke", Universität Stuttgart

Habermann, K. J.; Schittich, C. (1994) | Temporäre Roofing der antiken Arena in Nîmes (Temporary roofing of the ancient amphitheatre in Nîmes); in: Detail, Zeitschrift für Architektur + Baudetail, Institut für internationale Architektur-Dokumentation, München, p. 819 – 824

Heeg, M. (2007) | International Airport Bangkok – Engineering, Konfektion und Montage des Membrandachs (Suvarnabhumi International Airport Bangkok – engineering, fabrication and erection of the membrane roof); in: Zeitschrift Detail, Heft 7+8/2006 (Leichtbau + Systeme), Institut für internationale Architektur-Dokumentation, München, p. 824

Hemminger, R. (1990) | Preßklemmenverbindungen (Pressed clamp connections); in: Feyrer, K. u. a., Stehende Drahtseile und Seilendverbindungen, Expert-Verlag, Ehningen, p. 132 – 169

Herrmann, H. (1986) | Aufbereitungsanlagen (Preparation plant); in: Hensen, F.; Knappe, W.; Potente, H. (Hrsg.), Kunststoff-Extrusionstechnik II, Extrusionsanlagen, Carl Hanser Verlag, München, Vienna, p. 2 – 33

Holtermann, U. (2004) | Anforderungen beim Hochfrequenzschweißen beim Fügen von Textilien (Demands on high-frequency welding of joints in textiles); Vortrag zum Kolloquium "Konfektion technischer Textilien" am Institut für Textil- und Verfahrenstechnik, Stuttgart

Holst, S. (2006) | Suvarnabhumi International Airport Bangkok – Innovative Klimakonzeption (Suvarnabhumi International Airport Bangkok – innovative climatisation concept); in: Zeitschrift Detail, Heft 7 + 8/2006 (Leichtbau+Systeme), Institut für internationale Architektur-Dokumentation, München, p. 820

Imgrüth, H. (2002) | Beschrieb und Nachweise für das Arbeitsgerüst zur Montage der Kragenmembranen in Neuenburg (Description and design of the working scaffold for the cantilever membrane in Neuchâtel); Histec Engineering AG company material

Inauen, B. (2003) | Nicht veröffentlichtes, transkribiertes Tonbandprotokoll eines Interviews an der TU Vienna (Unpublished transcription of a taped interview at the Vienna Technical University)

Jeromin, W. (2003) | Gerüste und Schalungen im konstruktiven Ingenieurbau: Konstruktion und Bemessung (Scaffolds and formwork in structural engineering: design and detailing), Springer-Verlag, Heidelberg u. a.

Junker, D. (2004) | Nicht veröffentlichte Korrespondenz mit der Fa. VSL Schweiz AG (Unpublished correspondence with the company VSL Schweiz AG)

Kleinhanß, K. (1981) | Zur Montage biegeweicher Bauelemente (On the erection of flexible construction elements); in: Sonderforschungsbereich 64, Weitgespannte Flächentragwerke, Seile und Bündel im Bauwesen, Mitteilungen 59/81, Beratungsstelle für Stahlverwendung, Düsseldorf, p. III. 4 – 1 – III. 4 – 6

Klopfer, H. (1981) | Der Langzeitschutz von Zuggliedern (The long-term protection of tension members); in: Sonderforschungsbericht 64, Weitgespannte Flächentragwerke, Seile und Bündel im Bauwesen, Mitteilungen 59/81, Beratungsstelle für Stahlverwendung, Düsseldorf, p. II.4 – 1 – II.4 – 11

Labor Blum company material (1997) | Stuttgart, nicht veröffentlicht (Unpublished material from the Laboratory Blum)

Lenk, S. (2004) | Nicht veröffentlichtes, transkribiertes Tonbandprotokoll eines Interviews an der TU Vienna (Unpublished transcription of a taped interview at the Vienna Technical University)

Lenk, S. (2005) | Montage von Membranen (Erection of membranes); Vortrag an der TU-Vienna, Fakultät für Architektur und Raumplanung, Abteilung Hochbau 2, Jänner 2005

Lenk, S. (2006) | Membranmontage (Membrane erection); Vortrag an der TU-Vienna, Fakultät für Architektur und Raumplanung, Abteilung Hochbau 2, Dezember 2006

Leonhardt, F.; Schlaich, J. (1973) | Vorgespannte Seilnetzkonstruktionen. Das Olympiadach in München (Pretensioned rope net construction. The Olympic roof in Munich); in: Weitgespannte Flächentragwerke, Mitteilungen 19/1973, Sonderforschungsbericht 64, Universität Stuttgart

Lindner, J.; Breitschaft, G.; Thaten, J. (2001) | Die Seilkraftmessungen bei der Errichtung des Forumdaches beim Sony-Center am Potsdamer Platz in Berlin (The measurement of forces in ropes during the erection of the Forum roof of the Sony Center at the Potsdamer Platz in Berlin), Festschrift Sander, Berlin

Lindner, J.; Schulte, M.; Sischka, J.; Breitschaft, G.; Clarke, R.; Handel, E.; Zenkner, G. (1999) | Das Forumdach des Sony-Centers am Potsdamer Platz in Berlin (The Forum roof of the Sony Center at the Potsdamer Platz in Berlin); in: Stahlbau 68, Heft 12, Ernst & Sohn, Berlin, p. 975–994

Lorenz, T.; Mandl, P.; Siokola, W.; Zechner, M (2004) | Membranhülle am Flughafentower Vienna (Membrane envelope at the Vienna airport tower); in: Stahlbaurundschau, Oktober 2004, Österreichischer Stahlbauverband, Industriemagazin Verlag, Innsbruck, p. 48 – 51

Ludewig, S. (1974) | Montagebau – Grundlagen (Basics of construction erection), Wissensspeicher; VEB Verlag für Bauwesen, Berlin (Ost)

Miller, P. W. (2000) | Principles of Construction for Wide-Span Structures with Examples from the Millennium Dome; in: Widespan Roof Structures, Thomas Telford Publishing, London, Michael Barnes and Michael Dickson, p. 159–168

Minte, J. (1981) | Das mechanische Verhalten von Verbindungen beschichteter Chemiefasergewebe (The mechanical behaviour of joints in coated chemical fibre fabrics); Dissertation an der Fakultät für Maschinenwesen, RWTH Aachen

Mogk, R. (2000) | Bauen mit Seilen – Konstruktive Details aus der Praxis (Construction with ropes – design details from the practice); in: Leichte und ultraleichte Ingenieurbauten, 4. Dresdner Baustatik-Seminar, Lehrstuhl für Statik, Technische Universität Dresden, p. 141 – 157

Moncrieff, E.; Gründig, L.; Ströbel, D. (1999) | Zur Zuschnittsberechnung von Pilgerzelten bei der Phase II des Mina-Valley-Projektes (On the patterning of pilgrim tents for phase II of the Mina Valley Project); in: Bauen mit Textilien, Heft 4/1999, Deutscher Beton- und Bautechnik-Verein e. V., Berlin, p. 18 – 21

Mühlberger, H. (1984) | Darstellung und Klassifizierung von mechanisch gespannten Membrankonstruktionen nach Berechnung und Konstruktion zur Verwendung in bauaufsichtlichen Verfahren (Description and classification of mechanically tensioned membrane structures according to calculation and construction for use in building control procedures); Forschungsbericht, Institut für Bautechnik – IfBt Berlin (Förderer), IRB-Verlag, Stuttgart

Nentwing, J. (2000) | Kunststoff-Folien, Herstellung – Eigenschaften – Anwendung (Plastic foils, manufacture, properties, application); Carl Hanser Verlag, München, Vienna

Nolting-Firmenschrift (1999) | Temperatur nach Maß; Neuzeitliche Heiztechnik (Temperature made to measure; modern heating technology), Gustav Nolting GmbH

Oberbach, K. (2001) | Saechtling Kunststoff-Taschenbuch (Saechtling plastics pocket book), 28. Ausgabe, Carl Hanser Verlag, München, Vienna

Oplatka, G. (1983) | Das Vorrecken von Seilen (The prestretching of ropes); in: Internationale Seilbahn-Rundschau, 5/1983, p. 241 – 245

ÖNORM EN 12385-2 | Stahldrahtseile – Sicherheit – Teil 2: Begriffe, Bezeichnungen und Klassifizierung (Steel wire ropes – safety – part 2: terms, descriptions and classification), Ausgabe 2003 05 01, ON Österreichisches Normungsinstitut, Vienna, 2003

ÖNORM M 9500 | Stahldrahtseile – Sicherheit – allgemeine Bestimmungen (Steel wire ropes – safety – general regulations), Ausgabe 1980 09 01, ON Österreichisches Normungsinstitut, Vienna, 1980

Otto, F.; Happold, E.; Bubner, E. (1982) | IL 15, Lufthallenhandbuch (Air-supported hall handbook); Mitteilungen des Instituts für leichte Flächentragwerke, Karl Krämer Verlag, Stuttgart

Palkowski, S. (1990) | Statik der Seilkonstruktionen (Structural design of rope construction), Springer Verlag, Vienna u. a.

Peil, U. (2002) | Wire ropes – Seile – Herstellung und Eigenschaften (Ropes – manufacture and properties); in: Institut für Stahlbau, TU Braunschweig (Hrsg.), Bauen mit Seilen, Praxis-Seminar 2002, p. 1.1 – 1.21

Petzschmann, E.; Bauer, H. (1991) | Handbuch der Stahlbaumontage, Grundlagen für die Aus- und Weiterbildung des Montageführungspersonals (Steel erection handbook. rules for the initial and further training of erection foremen), Stahlbau-Verlagsgesellschaft mbH, Köln

PFEIFER company material (2003) | PFEIFER Holding GmbH & Co. KG, Memmingen, Germany

Philipp Holzmann – Baudokumentation (1988) | King Fahd, Internationales Stadion Riyadh (Construction documentation of the King Fahd International stadium in Riyadh), Filmproduktion: Peter Cürlis, Berlin; Auftraggeber: Ministerium für Jugend und Sport in Deutschland

Ramberger, G. (1978) | Die Berechnung der Normalkräfte in Zuggliedern über ihre Eigenfrequenz unter Berücksichtigung verschiedener Randbedingungen, der Biegesteifigkeit und der Dämpfung (The calculation of axial loads in tension members through their resonance frequency, taking into account various constraints, the bending stiffness and the attenuation); in: Stahlbau, Heft 10/78, Ernst & Sohn, Berlin, p. 314 – 318

Ramberger, G. (2003) | Nicht veröffentlichtes, transkribiertes Tonbandprotokoll eines Interviews am Institut für Stahlbau, TU Vienna (Unpublished transcription of a taped interview at the Institute for Steel Construction, Vienna Technical University)

Reitgruber, S. (2003) | Speichenradkonstruktionen (Spoked wheel construction), Diplomarbeit am Institut für Stahlbau, TU Vienna

Rudorf-Witrin, W. (1999) | Stichwortsammlung der Broschüre "50 Tipps – Der Weg zum Textilen Bauwerk", Herstellung und Montage (Keyword collection from the brochure "50 tips – the route to textile structure"), Frankfurt/M.

Rudorf-Witrin, W. (2004) | Hochfrequenzschweißen von Membranbauten (High-frequency welding of membrane buildings); Vortrag zum Kolloquium "Konfektion technischer Textilien" am Institut für Textil- und Verfahrenstechnik, Stuttgart

Rudorf-Witrin, W.; Stimpfle, B.; Blümel, S.; Pasternak, H. (2005) | Von der CargoLifter-Werfthalle zu Tropical Islands – Konstruktion, Herstellung und Montage der neuen ETFE-Folienkissen-Eindeckung (From the CargoLifter works hall to Tropical Islands – design detailing, production and erection of the new ETFE foil pillow external envelope) ; in: Bauingenieur, Band 81, Jänner 2006, Springer-VDI-Verlag, p. 33–42

Ryser, R.; Badoux, J.-C. (2002) | Ein avantgardistisches Bauwerk für die Expo.02 am Arteplage von Neuenburg (An avant-garde structure for the Expo.02 at the Arteplage in Neuchâtel) ; in: Stahlbau 71, Heft 8, Ernst & Sohn, Berlin p. 551 – 557

Saxe, K.; Kürten, R. (1992) | Zur Temperaturabhängigkeit des Kraft-Dehnungsverhalten PTFE-beschichteter Glasgewebe bei üblichen Verarbeitungstemperaturen (On the temperature-dependence of the force-strain behaviour of PTFE-coated fabrics at the usual processing temperatures); in: Bauingenieur 67, Springer-Verlag, p. 291 – 296

Scheffler, M. (1994) | Grundlagen der Fördertechnik – Elemente und Triebwerke (Basics of materials handling technology – elements and drives), Viehweg & Sohn Verlagsgesellschaft, Braunschweig, Wiesbaden

Schlaich, J.; Wagner, A. (1988) | Hybride Tragwerke 1+1=3. Das Ganze ist mehr als die Summe seiner Teile (Hybrid structures 1+1=3. The whole is more than the sum of its parts); in: Zeitschrift Baukultur, 6/88, p. 27 – 29

Schlaich, J.; Bergermann, R.; Göppert, K. (1999) | Textile Roofingen für die Sportstätten der Commonwealth Games 1998 in Kuala Lumpur/Malaysia (Textile roofing of the sports halls for the 1998 Commonwealth Games in Kuala Lumpur, Malaysia); in: Bauen mit Textilien, Heft 2/1999, Deutscher Beton- und Bautechnik-Verein e.V., Berlin, p. 13 – 19

Schlaich, M. (2000) | Kuppel und Kissen, Stierkampfarena-Dach in Madrid (Dome and pillow, bull fight arena roof in Madrid); in: Deutsche Bauzeitung, 9/2000, p. 59 – 69

Schwarz, O.; Ebeling, F.-W.; Furth, B. (1999) | Kunststoffverarbeitung (Plastics processing), Vogel Buchverlag, Würzburg

Seethaler, M. (2007) | Nicht veröffentlichte Korrespondenz mit der Fa. Hightex GmbH (Unpublished correspondence with the company Hightex GmbH)

Seliger, P. M. (1989) | Die Montage hoher Seilbahnstützen (The erection of high cableway pylons); Diplomarbeit am Institut für Eisenbahnwesen, TU Vienna

Singenstroth, F. (1998) | Drahtseilkonstruktionen, Definitionen; in (Wire rope construction, definitions): Laufende DrahtWire ropes: Bemessung und Überwachung (Running wire ropes: specification and monitoring), Feyrer, K. und 6 Mitautoren, Expert Verlag, Renningen-Malmsheim, p. 1 – 28

Siokola, W. (2004) | Nicht veröffentlichtes, transkribiertes Tonbandprotokoll eines Interviews bei der Fa. Zeman-Stahl, Vienna (Unpublished transcription of a taped interview at the company Zeman-Stahl, Vienna)

Sischka, J.; Stadler, F. (2003) | Das Forumdach des Sony-Centers in Berlin (The Forum roof of the Sony Center in Berlin); aus: http://www.waagner-biro.at/CDA/main/DE/SG/Referenzen/Membrankonstruktionen/Sony_Center.htm

Sobek, W. (1994) | Technologische Grundlagen des textilen Bauens (The technological basis of textile construction); in: Detail 6/1994, Institut für internationale Architektur – Dokumentation, München, p. 776 – 779

Sobek, W.; Speth, M. (1995) | Textile Werkstoffe (Textile materials); in: Bauingenieur 70, Springer-Verlag, p. 243 – 250

Sobek, W.; Linder, J.; Krampen, J. (2004) | Der neue Flughafen in Bangkok – eine ingenieurtechnische Herausforderung (The new airport in Bangkok – an engineering challenge); in: Stahlbau 73, Heft 7, Ernst & Sohn, Berlin, p. 461 – 467

Stauske, D. (1990) | Herstellung und Montage von Seilbauwerken (Production and erection of rope structures); in: Feyrer, K. u. a., Stehende Drahtseile und Seilendverbindungen, Expert-Verlag, Ehningen, p. 71 – 82

Stauske, D. (1995) | Drahtseile mit neuartigem Korrosionsschutz (Wire ropes with innovative corrosion protection); in: Techtextil-Symposium 1995, Frankfurt/M., Vortrag No. 533

Stauske, D. (2000) | Drahtseile für Seilbauwerke (Wire ropes for rope structures); in: Stahlbau, Heft 8/69, Ernst & Sohn, Berlin, p. 612 – 618

Stauske, D. (2002) | Seilbauwerke – Stehende Seile in der Architektur (Rope structures – static ropes in architecture); in: Kongressvortrag am 1. Internationalen Stuttgarter Seiltag, Institut für Fördertechnik und Logistik, Universität Stuttgart

Stavridis, L. (1992) | Seiltragwerke für große Spannweiten nach dem System Geiger (Rope structures for large spans according to the Geiger system); in: Bauingenieur 67, Springer-Verlag, p. 419 – 424

Steckelbach, C. (2005) | Nicht veröffentlichte Korrespondenz mit der Fa. Cenotec GmbH (Unpublished correspondence with the company Cenotec GmbH)

Stimpfle, B. (2000) | Leichte Flächentragwerke – Membrantragwerke (Lightweight structures – membrane structures); in: Leichte und ultraleichte Ingenieurbauten, 4. Dresdner Baustatik Seminar, Lehrstuhl für Statik, Technische Universität Dresden, p. 109 – 140

Stimpfle, B. (2004-1) | Luft formt – pneumatische Konstruktionen (Air forms – pneumatic construction); in: Kreative Ideen im Ingenieurbau, 8. Dresdner Baustatik-Seminar, Lehrstuhl für Statik, Technische Universität Dresden

Stimpfle, B. (2004-2) | Nicht veröffentlichtes, transkribiertes Tonbandprotokoll eines Interviews an der TU Vienna (Unpublished transcription of a taped interview at the Vienna Technical University)

Teschner, R. (2004) | Nicht veröffentlichte Korrespondenz mit dem Ingenieurbüro Teschner (Unpublished correspondence with the Teschner engineering consultancy)

Trurnit, P. D. (1981) | Zugglieder aus Stahldrähten, Herstellung und Eigenschaften (Tension members of wire rope, manufacture and properties); in: Sonderforschungsbericht 64, Weitgespannte Flächentragwerke Seile und Bündel im Bauwesen, Mitteilungen 59/81, Beratungsstelle für Stahlverwendung, Düsseldorf, p. I.2 – 1 – I.2 – 12

Verreet, R. (1996-1) | Über das Drehverhalten von Drahtseilen (About the twisting behaviour of wire ropes); in: Casar company material, PR GmbH Werbeagentur & Verlag GmbH, Aachen

Verreet, R. (1996-2) | Seilendverbindungen (Wire rope end connections); in: Casar company material, PR GmbH Werbeagentur & Verlag GmbH, Aachen

Verwaayen, J. (2002) | Grenzen der Drahtseilfertigung (Limits of wire rope production); in: Kongressvortrag am 1. Internationalen Stuttgarter Seiltag, Institut für Fördertechnik und Logistik, Universität Stuttgart

Vogel, W. (2002) | Bolzenverpressungen als Seilendverbindungen für sicherheitsrelevante Anwendungen (Swaged socket fittings to wire rope ends for safety-relevant applications); in: Kongressvortrag am 1. Internationalen Stuttgarter Seiltag, Institut für Fördertechnik und Logistik, Universität Stuttgart

Wegener, E. (2003) | Montagegerechte Anlagenplanung (Design of plant for ease of assembly), Wiley- VCH Verlag, Weinheim

Westerhoff, D. (1989) | Offene Spiralseile, Vollverschlossene Spiralseile (Open spiral strand, fully locked spiral strand); in: Thyssen company material, Thyssen-Draht AG

Zechner, M. (2005) | Städtebauliches Zeichen mit Signalcharakter (City planning sends a signal) ; in: [Umrisse] Zeitschrift für Baukultur, Heft 1/2005, Verlagsgruppe Wiederspahn, Wiesbaden, p. 24 – 27

Illustration acknowledgements

Fig. 2:	Praxishelfer Heben (Lifting practice guide) (1998), p. 27, Carl Stahl company material, reworked
Fig. 5:	*left:* Scheffler, M. (1994), p. 25
Fig. 7:	Scheffler, M. (1994), p. 27
Fig. 10:	Scheffler, M. (1994), p. 24
Fig. 11:	Scheffler, M. (1994), p. 24
Fig. 13:	*left, right:* Zeitschrift Draht 1/98, p. 37, p. 38, reworked
Fig. 14:	*left:* Westerhoff, D. (1989), p. 17 *right:* Casar-Spezialdrahtseile, Technische Eigenschaften (Technical properties of Casar wire ropes) (1997), p. 6, Casar company material
Fig. 15:	*left, right:* Casar-Spezialdrahtseile. Über das Drehverhalten von Seilen (1996), p. 2, p. 3, Casar company material
Fig. 16:	Seilbahnrundschau (Cableway news), 5/1981, p. 243
Fig. 17:	Stauske, D. (1990), p. 76, reworked
Fig. 18:	Full-Locked Coil Ropes for Bridges (1988), p. 14, Thyssen company material
Fig. 19:	Scheffler, M. (1994), p. 45
Fig. 20:	*left, right:* Feyrer, K. (1986), p. 102, p. 104
Fig. 21:	*left, middle, right:* Scheffler, M. (1994), p. 46
Fig. 22:	*left, middle:* Mogk, R. (2000), p. 154 *right:* Egger, G. (1999); Seile und Paralleldraht- bzw. Litzenbündel im Bauwesen (Ropes and parallel wires or strand bundles in construction); Diplomarbeit am Institut für Stahlbau, TU Vienna, p. 183
Fig. 23:	Casar-Spezialdrahtseile, Seilendverbindungen (1997), p. 3, Casar company material
Fig. 25:	*top:* Industrieseile (1991), p. 92, Thyssen company material *bottom left, bottom right:* Günter Ramberger, Vienna
Fig. 26:	http://www.bridon.de
Fig. 27:	based on Hemminger, R. (1990), p. 132
Fig. 28:	*left:* Hemminger, R. (1990), p. 136 *middle, right:* Casar-Seilendverbindungen (1997), p. 33, p. 36
Fig. 29:	Mogk, R. (2000), p. 148
Fig. 30:	Casar-Seilendverbindungen (1997), p. 46
Fig. 31:	*top:* Casar-Seilendverbindungen (1997), p. 48 *middle:* Hemminger, R. (1990), p. 166 *bottom:* Casar-Seilendverbindungen (1997), p. 48
Fig. 32:	*left:* Hemminger, R. (1990), p. 180 *right:* Casar-Seilendverbindungen (1997), p. 5
Fig. 33:	*left:* Mogk, R. (2000), p. 156
Fig. 34:	*left:* Mogk, R. (2000), p. 156 *middle:* Stauske, D. (1990), p. 80 *right:* Montageservice SL GmbH, Stefan Lenk, Hallbergmoos
Fig. 35:	*left,* middle: Zeman GmbH, Walter Siokola, Vienna *right:* Architekturbüro Silja Tillner, Vienna
Fig. 36:	*left:* Forum Dach Sony Berlin, Potsdamer Platz; Waagner-Biro company material *right:* Inauen-Schätti AG, Bruno Inauen, Schwanden
Fig. 38:	*top, bottom:* Bandwebmaschine NG3 (2003), Jakob Müller company material
Fig. 39:	Covertex GmbH, Obing
Fig. 42:	IF, Ingenieurgemeinschaft Flächentragwerke, Horst Dürr, Reichenau
Fig. 45:	*left,* middle: Burkhard, W. (1998), Textile Fertigungsverfahren (Textile production processes), Hanser Verlag, München, p. 35, p. 56 *right:* Fritz, C. P. (1999)
Fig. 46:	*left:* Blum, R. (1990), p. 212
Fig. 48:	*right:* Dornier-Greiferwebmaschine Typ PS, p. 7, Dornier company material
Fig. 50:	*left, right:* Montageservice SL GmbH, Stefan Lenk, Hallbergmoos
Fig. 53:	*left, right:* IF, Ingenieurgemeinschaft Flächentragwerke, Horst Dürr, Reichenau
Fig. 54:	*left, right:* IF, Ingenieurgemeinschaft Flächentragwerke, Horst Dürr, Reichenau
Fig. 55:	*top, bottom:* Baumann, Th. (2002), p. 159
Fig. 56:	*top left:* Kaltenbach, F. (Hg.) (1994), Detail-Band Praxis, Transluzente Materialien (Detail volume practice, translucent materials), Institut für Internationale Architektur-Dokumentation, München, p. 68 *top right, bottom:* Interglas Technologies AG, Erbach
Fig. 57:	*left, middle, right:* Domininghaus, H. (1992), p. 9
Fig. 58:	*top:* Kaltenbach, F. (Hg.) (1994), Detail-Band Praxis, p. 69
Fig. 59:	*left, right:* Ebnesajjad, p. 44 (2000); Fluoroplastics, Volume 1: Non-Melt Processible Fluoroplastics, William Andrew Publishing/Plastics Design Library
Fig. 60:	Schwarz, O.; Ebeling, F.-W.; Furth, B. (1999), p. 19
Fig. 61:	*left, right:* Nentwing, J. (2000), p. 49, p. 53
Fig. 62:	*top left, bottom left, right:* Schwarz, O.; Ebeling, F.-W.; Furth, B. (1999), p. 52, p. 52, p. 41
Fig. 63:	Bongaerts, H. (1986), in: Hensen, F.; Knappe, W.; Potente, H. (Hg.), Kunststoff-Extrusionstechnik II, Extrusionsanlagen (Plastic extrusion technology II, extrusion plant), Carl Hanser Verlag, München, Vienna, p. 147
Fig. 64:	Nowoflon ET®, Technische Informationen, Nowofol company material, Siegsdorf 2004, p. 10
Fig. 66:	Blum, R. (1990), p. 205, reworked
Fig. 67:	Blum, R. (1990), p. 204, reworked
Fig. 68:	Blum, R. (1990), p. 212
Fig. 71:	*right:* Systemleichtbau I: Mechanisch vorgespannte Membranen (Lightweight system construction I: mechanically pretensioned membranes); Vorlesungen am Institut für Leichtbau Enwerfen und Konstruieren, Universität Stuttgart, p. 5
Fig. 72:	*left:* Blum, R. (1990), p. 219 *right:* Test carried out at Labor Blum, received from: IF, Ingenieurgemeinschaft Flächentragwerke, Reichenau
Fig. 73:	Test carried out at Labor Blum, received from: IF, Ingenieurgemeinschaft Flächentragwerke, Reichenau
Fig. 80:	*left:* Forum roof Sony Berlin, Potsdamer Platz, Waagner-Biro company material
Fig. 82:	*left, right:* Hellerich, W.; Harsch, G.; Haenle, S. (2001), p. 344, reworked
Fig. 83:	*left, right:* Hellerich, W.; Harsch, G.; Haenle, S. (2001), p. 33
Fig. 84:	*left, right:* Minte, J. (1981), p. 27, p. 28
Fig. 93:	Bauen mit Textilien (Textile construction), Heft 2/1999, p. 18

Fig. 100:	*left, right:* Montageservice SL GmbH, Stefan Lenk, Hallbergmoos
Fig. 101:	*left, middle, right:* Montageservice SL GmbH, Stefan Lenk, Hallbergmoos
Fig. 104:	*left:* Montageservice SL GmbH, Stefan Lenk, Hallbergmoos *right:* project, implementation and photo © Skyspan (Europe) GmbH
Fig. 107:	Zünd company material (2004)
Fig. 108:	*right:* Magnus Malin NHZ GmbH
Fig. 109:	Sobek, W.; Speth, M. (1995), p. 248, reworked
Fig. 110:	*left:* Holtermann U. (2004), reworked middle: Fügetechnik (Joint technology); DVS-Fachgruppe 8.1, p. 184, reworked
Fig. 111:	a – d: Ceno Tec GmbH, Wolfgang Rudorf-Witrin, Greven
Fig. 112:	Ceno Tec GmbH, Wolfgang Rudorf-Witrin, Greven
Fig. 113:	*left, middle, right:* Ceno Tec GmbH, Wolfgang Rudorf-Witrin, Greven
Fig. 114:	Ceno Tec GmbH, Wolfgang Rudorf-Witrin, Greven
Fig. 116:	*left, right:* Covertex GmbH, Obing
Fig. 117:	*left, right:* Eichenhofer GmbH, Illertissen
Fig. 118:	*left:* Birdair, Inc., Amherst *right:* Schlaich Bergermann & Partner GmbH, Stuttgart
Fig. 123:	Montageservice SL GmbH, Stefan Lenk, Hallbergmoos
Fig. 125:	*left:* IF, Ingenieurgemeinschaft Flächentragwerke, Horst Dürr, Reichenau
Fig. 128:	*left, right:* Detail, 8/1996, p. 1243, cover illustration, reworked
Fig. 130:	Rein, A; Wilhelm, V. (2000), Detail, No. 6, p. 1047, reworked; und: Sobek, W.; Speth, M. (1995), p. 248, reworked
Fig. 134:	*left:* Stahl und Licht, Das Dach des Sony Center am Potsdamer Platz, Fotografien von Roland Horn (Steel and light, the roof of the Sony Center in Berlin, photography by Roland Horn) (2000), Nicolaische Verlagsbuchhandlung Beuermann GmbH, Berlin, p. 89
Fig. 137:	Gropper, H.; Sobek, W. (1985), Tagungsband SFB-64, Hans Gropper, Werner Sobek, p. 7
Fig. 146:	*left:* Covertex GmbH, Obing *right:* formTL, Ingenieure für Tragwerk und Leichtbau GmbH, Radolfzell
Fig. 147:	*left:* formTL, Ingenieure für Tragwerk und Leichtbau GmbH, Radolfzell
Fig. 152:	PFEIFER company material
Fig. 153:	Ceno Tec GmbH, Wolfgang Rudorf-Witrin, Greven
Fig. 154:	IF, Ingenieurgemeinschaft Flächentragwerke, Horst Dürr, Reichenau
Fig. 155:	IF, Ingenieurgemeinschaft Flächentragwerke, Horst Dürr, Reichenau
Fig. 156:	IPL, Ingenieurplanung Leichtbau GmbH, Radolfzell
Fig. 157:	*left, right:* Phillip Holzmann AG, Saudia Arabia
Fig. 161:	*left, right:* IPL Ingenieurplanung Leichtbau GmbH, Radolfzell
Fig. 163:	*left:* Drees, G.; Krauß, S. (2002), p. 35 *right:* Liebherr company material
Fig. 164:	*left:* Liebherr company material *right:* Drees, G.; Krauß, S. (2002), p. 38
Fig. 165:	Wegener, E. (2003), p. 305 – 306
Fig. 166:	Drees, G.; Krauß, S. (2002), p. 45
Fig. 167:	based on: Ludewig, S. (1974), p. 404
Fig. 168:	*left:* Montageservice SL GmbH, Stefan Lenk, Hallbergmoos *right:* Christo and Jeanne-Claude, Wrapped Reichstag, Berlin 1971 – 95, Taschen Verlag, Köln 2001; Fotos: Wolfgang Volz
Fig. 169:	Meili company material, reworked
Fig. 170:	*left:* Grundlagen der Fördertechnik (Basics of materials handling technology), Vieweg, 1994 *right:* Inauen-Schätti AG, Bruno Inauen, Schwanden
Fig. 171:	*left:* Waagner-Biro company material *right:* Eberspächer company material
Fig. 172:	*left, right:* Inauen-Schätti AG, Bruno Inauen, Schwanden Fig. 173: *left, right:* Ferjencik, P.; Tochacek, M. (1975), p. 115, reworked
Fig. 174:	*left, right:* Montageservice SL GmbH, Stefan Lenk, Hallbergmoos
Fig. 175:	Eberspächer company material
Fig. 176:	*left:* VSL (Switzerland) Ltd., Daniel Junker, Subingen *middle, right:* Umbaudokumentation Neues Frankfurter Waldstadion (Documentation of the rebuilding of the new Waldstadion in Frankfurt): © fantasticweb new media GmbH
Fig. 177:	Inauen-Schätti AG, Bruno Inauen, Schwanden
Fig. 178:	Inauen-Schätti AG, Bruno Inauen, Schwanden
Fig. 179:	*left, right:* architektur.aktuell, 3/2004, p. 88, p. 81
Fig. 180:	*right:* Stahl und Licht, Das Dach des Sony Center am Potsdamer Platz, Fotografien von Roland Horn (Steel and light, the roof of the Sony Center in Berlin, photography by Roland Horn) (2000), Nicolaische Verlagsbuchhandlung Beuermann GmbH, Berlin, p. 62
Fig. 181:	*left:* Meili company material *right:* Montageservice SL GmbH, Stefan Lenk, Hallbergmoos
Fig. 182:	Montageservice SL GmbH, Stefan Lenk, Hallbergmoos
Fig. 183:	Vetter company material
Fig. 184:	*left, right:* PFEIFER company material 2003, reworked
Fig. 185:	*left:* IF, Ingenieurgemeinschaft Flächentragwerke, Horst Dürr, Reichenau *right:* PFEIFER company material 2003
Fig. 187:	*left:* Montageservice SL GmbH, Stefan Lenk, Hallbergmoos *right:* still from video – Leichtes Spiel der Kräfte (Light play of forces), PFEIFER Seil- und Hebetechnik GmbH, Memmingen
Fig. 188:	*middle, right:* Montageservice SL GmbH, Stefan Lenk, Hallbergmoos
Fig. 189:	*left, right:* still from video, PFEIFER Seil- und Hebetechnik GmbH, Memmingen
Fig. 190:	IF, Ingenieurgemeinschaft Flächentragwerke, Horst Dürr, Reichenau
Fig. 194:	*left, right:* Montageservice SL GmbH, Stefan Lenk, Hallbergmoos
Fig. 195:	*left:* Schlaich Bergermann & Partner GmbH, Stuttgart *right:* Architekturbüro Silja Tillner, Vienna
Fig. 196:	*left:* Montageservice SL GmbH, Stefan Lenk, Hallbergmoos
Fig. 197:	*left, right:* Birdair, Inc., Amherst
Fig. 198:	*left:* Montageservice SL GmbH, Stefan Lenk, Hallbergmoos
Fig. 199:	*left, right:* Histec Engineering AG, Hansruedi Imgrüth, Buochs
Fig. 200:	*right:* Montageservice SL GmbH, Stefan Lenk, Hallbergmoos
Fig. 201:	*left:* Schlaich Bergermann & Partner GmbH, Stuttgart *right:* Birdair, Inc., Amherst
Fig. 202:	Montageservice SL GmbH, Stefan Lenk, Hallbergmoos
Fig. 203:	*top:* Günter Ramberger, Vienna *bottom:* Zeman GmbH, Walter Siokola, Vienna
Fig. 204:	*left, right:* Umbaudokumentation Neues Frankfurter Waldstadion (Documentation of the rebuilding of the new Waldstadion in Frankfurt): © fantasticweb new media GmbH
Fig. 206:	a, b, d: Montageservice SL GmbH, Stefan Lenk, Hallbergmoos c: Umbaudokumentation Neues Frankfurter Waldstadion (Documentation of the rebuilding of the new Waldstadion in Frankfurt): © fantasticweb new media GmbH
Fig. 207:	a, b, c: Montageservice SL GmbH, Stefan Lenk, Hallbergmoos d: Inauen-Schätti AG, Bruno Inauen, Schwanden

Fig. 209:	*top:* Birdair, Inc., Amherst *bottom:* Stahl und Licht, Das Dach des Sony Center am Potsdamer Platz, Fotografien von Roland Horn (Steel and light, the roof of the Sony Center in Berlin, photography by Roland Horn) (2000), Nicolaische Verlagsbuchhandlung Beuermann GmbH, Berlin, p. 73
Fig. 213:	*left:* Montageservice SL GmbH, Stefan Lenk, Hallbergmoos *right:* Stahl und Licht, Das Dach des Sony Center am Potsdamer Platz, Fotografien von Roland Horn (Steel and light, the roof of the Sony Center in Berlin, photography by Roland Horn) (2000), Nicolaische Verlagsbuchhandlung Beuermann GmbH, Berlin, p. 66
Fig. 215:	*left:* Montageservice SL GmbH, Stefan Lenk, Hallbergmoos *middle:* still from video, PFEIFER Seil- und Hebetechnik GmbH, Memmingen *right:* Inauen-Schätti AG, Bruno Inauen, Schwanden
Fig. 216:	*left:* Zeman GmbH, Walter Siokola, Vienna *middle:* Montageservice SL GmbH, Stefan Lenk, Hallbergmoos *right:* Christo and Jeanne-Claude, Wrapped Reichstag, Berlin
Fig. 218:	*left:* formTL, Ingenieure für Tragwerk und Leichtbau GmbH, Radolfzell *middle:* Montageservice SL GmbH, Stefan Lenk, Hallbergmoos
Fig. 219:	Birdair, Inc., Amherst
Fig. 220:	formTL, Ingenieure für Tragwerk und Leichtbau GmbH, Radolfzell
Fig. 221:	Barnes M., Dickson, M. (2000); Widespan Roof Structures, Thomas Telford Publishing, London, p. 160
Fig. 224:	Schlaich Bergermann & Partner GmbH, Stuttgart
Fig. 225:	Schlaich Bergermann & Partner GmbH, Stuttgart
Fig. 226:	*left:* Stahl und Licht, Das Dach des Sony Center am Potsdamer Platz, Fotografien von Roland Horn (Steel and light, the roof of the Sony Center in Berlin, photography by Roland Horn) (2000), Nicolaische Verlagsbuchhandlung Beuermann GmbH, Berlin, p. 55 *middle:* Forum roof Sony Berlin, Potsdamer Platz; Waagner-Biro company material *right:* Lindner, J.; Schulte, M.; Sischka, J.; Breitschaft, G.; Clarke, R.; Handel, E.; Zenkner, G. (1999); p. 984
Fig. 227:	*left:* Lindner, J.; Schulte, M.; Sischka, J.; Breitschaft, G.; Clarke, R.; Handel, E.; Zenkner, G. (1999); p. 980 *middle:* Stahl und Licht, Das Dach des Sony Center am Potsdamer Platz, Fotografien von Roland Horn (Steel and light, the roof of the Sony Center in Berlin, photography by Roland Horn) (2000), Nicolaische Verlagsbuchhandlung Beuermann GmbH, Berlin, p. 20 *right:* Stahl und Licht, Das Dach des Sony Center am Potsdamer Platz, Fotografien von Roland Horn (2000), Nicolaische Verlagsbuchhandlung Beuermann GmbH, Berlin, p. 113
Fig. 228:	*left:* Zeman company material *right:* Stahlbau Rundschau (Steel construction news), Oktober 2004, p. 48
Fig. 229:	*left, middle, right:* Zeman GmbH, Walter Siokola, Vienna
Fig. 230:	*left:* Ingenieurbüro Karlheinz Wagner, Vienna *right:* Photo: © Werner Kaligofsky, Vienna
Fig. 231:	*left:* Ingenieurbüro Karlheinz Wagner, Vienna *top right:* Jabornegg & Pálffy Architekten, Vienna *bottom right:* RW Tragwerksplanung, Vienna
Fig. 232:	*left,* middle: Ingenieurbüro Karlheinz Wagner, Vienna
Fig. 236:	Stavridis, L. (1992); p. 420
Fig. 238:	Schlaich Bergermann & Partner GmbH, Stuttgart, reworked
Fig. 239:	*left, middle, right:* still from video, PFEIFER Seil- und Hebetechnik GmbH, Memmingen
Fig. 240:	*left, right:* Bauingenieur 70, (1995), Springer-Verlag, p. 258, Düsseldorf
Fig. 243:	*left, middle, right:* Tensoforma Trading, Bergamo
Fig. 245:	*left, middle, right:* Montageservice SL GmbH, Stefan Lenk, Hallbergmoos
Fig. 247:	*left, middle, right:* Schlaich Bergermann & Partner GmbH, Stuttgart
Fig. 248:	Montageservice SL GmbH, Stefan Lenk, Hallbergmoos
Fig. 249:	Montageservice SL GmbH, Stefan Lenk, Hallbergmoos
Fig. 250:	Montageservice SL GmbH, Stefan Lenk, Hallbergmoos
Fig. 251:	Montageservice SL GmbH, Stefan Lenk, Hallbergmoos
Fig. 252:	Montageservice SL GmbH, Stefan Lenk, Hallbergmoos
Fig. 253:	Montageservice SL GmbH, Stefan Lenk, Hallbergmoos
Fig. 255:	*left, middle, right:* Montageservice SL GmbH, Stefan Lenk, Hallbergmoos
Fig. 257:	*left, right:* Inauen-Schätti AG, Bruno Inauen, Schwanden
Fig. 258:	Montageservice SL GmbH, Stefan Lenk, Hallbergmoos
Fig. 259:	Montageservice SL GmbH, Stefan Lenk, Hallbergmoos
Fig. 260:	Montageservice SL GmbH, Stefan Lenk, Hallbergmoos
Fig. 261:	Montageservice SL GmbH, Stefan Lenk, Hallbergmoos
Fig. 262:	Montageservice SL GmbH, Stefan Lenk, Hallbergmoos
Fig. 263:	Montageservice SL GmbH, Stefan Lenk, Hallbergmoos
Fig. 265:	4: Schlaich Bergermann & Partner GmbH, Stuttgart, reworked
Fig. 266:	*left:* Architekturbüro Silja Tillner, Vienna *middle, right:* Birdair, Inc., Amherst
Fig. 268:	1, 2: Montageservice SL GmbH, Stefan Lenk, Hallbergmoos
Fig. 270:	*left, middle, right:* Montageservice SL GmbH, Stefan Lenk, Hallbergmoos
Fig. 271:	Hightex GmbH
Fig. 272:	Werner Sobek Ingenieure, Stuttgart
Fig. 273:	Hightex GmbH
Fig. 274:	Hightex GmbH
Fig. 275:	Werner Sobek Ingenieure, Stuttgart
Fig. 276:	Werner Sobek Ingenieure, Stuttgart
Fig. 278:	*left, middle, right:* Montageservice SL GmbH, Stefan Lenk, Hallbergmoos
Fig. 279:	Phillip Holzmann AG, Saudia Arabia
Fig. 281:	Ceno Tec GmbH, Greven
Fig. 282:	formTL, Ingenieure für Tragwerk und Leichtbau GmbH, Radolfzell
Fig. 283:	formTL, Ingenieure für Tragwerk und Leichtbau GmbH, Radolfzell
Fig. 284:	Montageservice SL GmbH, Stefan Lenk, Hallbergmoos
Fig. 285:	Montageservice SL GmbH, Stefan Lenk, Hallbergmoos
Fig. 286:	Montageservice SL GmbH, Stefan Lenk, Hallbergmoos
Fig. 288:	*top:* www.reber-montagen.ch *bottom:* Inauen-Schätti AG, Bruno Inauen, Schwanden
Fig. 289:	*left:* Umbaudokumentation Neues Frankfurter Waldstadion (Documentation of the rebuilding of the new Waldstadion in Frankfurt): © fantasticweb new media GmbH *middle:* still from video, PFEIFER Seil- und Hebetechnik GmbH, Memmingen *right:* Günter Ramberger
Fig. 290:	*left:* Inauen-Schätti AG, Bruno Inauen, Schwanden *right:* still from video, PFEIFER Seil- und Hebetechnik GmbH, Memmingen
Fig. 291:	*left:* Inauen-Schätti AG, Bruno Inauen, Schwanden *middle, right:* Umbaudokumentation Neues Frankfurter Waldstadion: © fantasticweb new media GmbH

Fig. 292:	*left:* PFEIFER Seil- und Hebetechnik GmbH, Memmingen *right:* Umbaudokumentation Neues Frankfurter Waldstadion: © fantasticweb new media GmbH	Fig. 322:	Deutsche BauZeitschrift, 4/2003
Fig. 293:	formTL, Ingenieure für Tragwerk und Leichtbau GmbH, Radolfzell	Fig. 323:	Deutsche BauZeitschrift, 4/2003
Fig. 294:	*left:* IF, Ingenieurgemeinschaft Flächentragwerke, Horst Dürr, Reichenau *right:* Birdair, Inc., Amherst	Fig. 324:	Leicht Weit – Light Structures, Jörg Schlaich and Rudolf Bergermann, Ausstellungskatalog Deutsches Architekturmuseum Frankfurt am Main, Prestel Verlag, 2004
Fig. 295:	*left:* Christo and Jeanne-Claude, Wrapped Reichstag, Berlin *right:* Montageservice SL GmbH, Stefan Lenk, Hallbergmoos	Fig. 325:	Deutsche BauZeitschrift, 4/2003
Fig. 296:	*left, right:* Montageservice SL GmbH, Stefan Lenk, Hallbergmoos	Fig. 326:	Schlaich Bergermann & Partner GmbH, Stuttgart
Fig. 299:	*left, right:* Montageservice SL GmbH, Stefan Lenk, Hallbergmoos	Fig. 327:	*left, right:* IF, Ingenieurgemeinschaft Flächentragwerke, Horst Dürr, Reichenau
Fig. 300:	*left:* still from video, PFEIFER Seil- und Hebetechnik GmbH, Memmingen *right:* Günter Ramberger, Vienna	Fig. 328:	www.if-group.de; IF, Ingenieurgemeinschaft Flächentragwerke, Reichenau
Fig. 301:	*left:* Umbaudokumentation Neues Frankfurter Waldstadion (Documentation of the rebuilding of the new Waldstadion in Frankfurt): © fantasticweb new media GmbH *right:* PFEIFER Seil- und Hebetechnik GmbH, Memmingen, Jürgen Winkler	Fig. 329:	*left, right:* IF, Ingenieurgemeinschaft Flächentragwerke, Horst Dürr, Reichenau
		Fig. 330:	*left, middle:* Ryser, R.; Badoux, J.-C. (2002); p. 555 *right:* www.tensinet.com
Fig. 302:	*left:* Barnes M., Dickson, M. (2000); Widespan Roof Structures, Thomas Telford Publishing, London, p. 165 *middle, right:* Inauen-Schätti AG, Bruno Inauen, Schwanden	Fig. 331:	IF, Ingenieurgemeinschaft Flächentragwerke, Horst Dürr, Reichenau, reworked
		Fig. 332:	Ryser, R.; Badoux, J.-C. (2002); p. 556, reworked
Fig. 303:	*left, right:* IF, Ingenieurgemeinschaft Flächentragwerke, Horst Dürr, Reichenau	Fig. 333:	*left, right:* Montageservice SL GmbH, Stefan Lenk, Hallbergmoos
Fig. 304:	*left, right:* formTL, Ingenieure für Tragwerk und Leichtbau GmbH, Radolfzell	Fig. 342:	*left, right:* Covertex GmbH, Obing
Fig. 306:	*left, right:* Montageservice SL GmbH, Stefan Lenk, Hallbergmoos	Fig. 343:	*left, right:* Birdair, Inc., Amherst
Fig. 307:	*left, right:* Montageservice SL GmbH, Stefan Lenk, Hallbergmoos	Fig. 344:	*right:* Montageservice SL GmbH, Stefan Lenk, Hallbergmoos, reworked
Fig. 308:	*left, right:* Christo and Jeanne-Claude, Wrapped Reichstag, Berlin	Fig. 346:	*left, right:* Montageservice SL GmbH, Stefan Lenk, Hallbergmoos
Fig. 309:	*left:* Montageservice SL GmbH, Stefan Lenk, Hallbergmoos *middle:* Forum-Dach Sony-Berlin, Potsdamer Platz; Waagner-Biro company material *right:* Stahl und Licht, Das Dach des Sony Center am Potsdamer Platz, Fotografien von Roland Horn (Steel and light, the roof of the Sony Center in Berlin, photography by Roland Horn) (2000), Nicolaische Verlagsbuchhandlung Beuermann GmbH, Berlin, p. 70	Fig. 347:	*left, right:* Projekt, Realisation und Foto © Skyspan (Europe) GmbH
		Fig. 349:	IF, Ingenieurgemeinschaft Flächentragwerke, Horst Dürr, Reichenau
		Fig. 350:	*left, right:* Montageservice SL GmbH, Stefan Lenk, Hallbergmoos
		Fig. 352:	*left, middle, right:* Montageservice SL GmbH, Stefan Lenk, Hallbergmoos
Fig. 310:	*left, right:* Inauen-Schätti AG, Bruno Inauen, Schwanden	Fig. 354:	Montageservice SL GmbH, Stefan Lenk, Hallbergmoos
Fig. 311:	*left:* Ingenieurbüro Teschner, Rochus Teschner, Kosel *right:* Umbaudokumentation Neues Frankfurter Waldstadion (Documentation of the rebuilding of the new Waldstadion in Frankfurt): © fantasticweb new media GmbH	Fig. 355:	*left, right:* Montageservice SL GmbH, Stefan Lenk, Hallbergmoos
		Fig. 360:	*left, right:* Schlaich Bergermann & Partner GmbH, Stuttgart
		Fig. 361:	*left, right:* Schlaich Bergermann & Partner GmbH, Stuttgart
Fig. 312:	Leicht Weit – Light Structures, Jörg Schlaich and Rudolf Bergermann, Ausstellungskatalog Deutsches Architekturmuseum Frankfurt am Main, Prestel Verlag, 2004	Fig. 363:	*left:* Montageservice SL GmbH, Stefan Lenk, Hallbergmoos *right:* IF, Ingenieurgemeinschaft Flächentragwerke, Reichenau
		Fig. 364:	*left:* Montageservice SL GmbH, Stefan Lenk, Hallbergmoos *right:* IF, Ingenieurgemeinschaft Flächentragwerke, Reichenau
Fig. 313:	*left:* IF, Ingenieurgemeinschaft Flächentragwerke, Horst Dürr, Reichenau *right:* Detail 6/1994, p. 822	Fig. 365:	*left, right:* IF, Ingenieurgemeinschaft Flächentragwerke, Horst Dürr, Reichenau
Fig. 314:	IF, Ingenieurgemeinschaft Flächentragwerke, Horst Dürr, Reichenau	Fig. 366:	Blum, R. (2000), reworked
Fig. 315:	IF, Ingenieurgemeinschaft Flächentragwerke, Horst Dürr, Reichenau	Fig. 367:	Montageservice SL GmbH, Stefan Lenk, Hallbergmoos, reworked
Fig. 316:	IF, Ingenieurgemeinschaft Flächentragwerke, Horst Dürr, Reichenau	Fig. 368:	RTM 20D Seilkraftmesser, PIAB company material
Fig. 317:	1 – 10: IF, Ingenieurgemeinschaft Flächentragwerke, Horst Dürr, Reichenau	Fig. 369:	*left, right:* Blum, R. (1982); p. 271, p. 272
Fig. 318:	IF, Ingenieurgemeinschaft Flächentragwerke, Horst Dürr, Reichenau	Fig. 370:	*right:* Montageservice SL GmbH, Stefan Lenk, Hallbergmoos
Fig. 319:	Structural Engineering Review Band 6, No. 3 – 4, Elsevier Science, 1994, p. 211 – 213	Fig. 371:	Labor-Blum company material
Fig. 320:	a: IF, Ingenieurgemeinschaft Flächentragwerke, Horst Dürr, Reichenau b – f: Structural Engineering Review Band. 6, No. 3 – 4, Elsevier Science, 1994, p. 210		
Fig. 321:	Deutsche BauZeitschrift, 4/2003		

Figures not listed have been provided by the author.

Projects (1989–2007)

01 | Forum roof, Sony Center Berlin / Germany

Location:	Central Berlin / Germany
Completion:	2000
Membrane area:	approx. 5,800 m²
Membrane material type:	Glass/PTFE fabric
Client:	SONY, Kajima, Tishman Speyer, New York-Berlin
Architect:	Murphy/Jahn, Chicago / USA
Structural engineer:	Ove Arup & Partners, New York / USA
Erection planning:	Zenkner & Handel, Graz / Austria
Steel erection:	Waagner Biro, Vienna / Austria
Wire ropes:	PFEIFER, Memmingen / Germany
Membrane fabrication:	Birdair, Amherst / USA
Membrane erection:	Birdair, Amherst / USA
Photo:	© Roland Horn, Berlin / Germany

02 | Atrium roofing, Forum Kirchberg / Luxembourg

Location:	Kirchberg, Luxembourg / Luxembourg
Completion:	1997
Roofed area:	approx. 2,700 m²
Membrane material type:	Glass/PTFE fabric
Client:	Forum Kirchberg s.a., Auchan, Luxembourg / Luxembourg
Architect:	Martin Lammar (atelier a+u) Luxembourg in partnership with Lars Iwdal (arkitektbyran ab) Göteborg / Sweden and in collaboration with Murray Church (HT-Lux) Luxembourg
General contractor for membrane roof:	CENO TEC, Greven / Germany
Structural design and patterning:	IPL, Radolfzell / Germany
Steel erection:	Steel erection Zwickau, Zwickau / Germany
Wire ropes:	PFEIFER, Memmingen / Germany
Membrane fabrication:	CENO TEC, Greven / Germany
Membrane erection:	CENO TEC, Greven / Germany
Photo:	© CENO TEC, Greven / Germany

ETFE-Roof – Oldtimer-Museum Düsseldorf

Intercity Railway Station, Airport Leipzig

Grandstand Roofing Stadion AL Ain, VAE

ETFE-Roof – Hitachi Office, Duisburg

Amusement Sierksdorf

Gerry-Weber-Stadion, Halle

Textile Light Construction from CENO TEC: Traditional and future orientated

The long-established company CENO TEC is one the most renown businesses in the area of textile architecture realising challenging membrane constructions world wide.

Architecture-membranes made of high performance textile fabrics and ETFE foils are ideal building materials – in connection with steel supporting structures and steel ropes it is of no problem to realise highly resilient and yet filigree roof and facade constructions. The exceptional degree of aesthetics and translucency extinguishes the ultimate fascination for this type of architecture. The satisfying short term planning and erection periods, sustained durability and cost-efficient realisations make these constructions extremely convincing.

CENO TEC accommodates as Full-Service-Contractor a highly qualified completion – beginning with the project planning going on to the construction design and production and ending with the final on-site erection.

CENO TEC GmbH

Am Eggenkamp 14
D – 48268 Greven
Telefon 02571 969-0
Fax 02571 3300
E-Mail info@ceno-tec.de

www.ceno-tec.de

03 | Roofing, EKZ Grossfeldsiedlung, Vienna / Austria

Location:	Vienna Floridsdorf / Austria
Completion:	construction phase 1: 2005; phase 2: 2007
Area:	approx. 3,500 m² (construction phases 1 und 2)
Membrane material type:	Glass/PTFE fabric, glass mesh fabric
Client:	Ekazent Immobilien Management GmbH, Vienna / Austria
General contractor:	DELTA Projektconsult, Vienna / Austria
Architect:	DELTA Projektconsult, Vienna / Austria
Structural engineer:	Teschner, Kosel / Germany
Design of patterning:	Teschner, Kosel / Germany
Design of steelwork:	Teschner, Kosel / Germany
Fire protection design:	Ingenieurbüro für Brandschutz und Sicherheit Düh, Vienna / Austria
Steelwork and erection:	Profilstahl, Judenburg / Austria
Membrane fabrication:	Cenotec, Greven / Germany
Membrane erection:	Velabran, München / Germany
Photo:	© Michael Seidel, Vienna / Austria

04 | Roofing, Urban Loritz Platz, Vienna / Austria

Location:	Vienna / Austria
Completion:	1999
Roofed area:	approx. 2,000 m²
Membrane material type:	PES/PVC fabric
Client:	Magistrat der Stadt Vienna / Austria
Architect:	Silja Tillner, Vienna / Austria
Structural engineer:	Schlaich Bergermann & Partner, Stuttgart / Germany
Erection planning:	Schlaich Bergermann & Partner, Stuttgart / Germany
Steel erection:	Stuag Bau, Vienna / Austria
Supply of wire ropes:	PFEIFER, Memmingen / Germany
Membrane fabrication:	Skyspan Europe, Rimsting / Germany
Membrane erection:	Skyspan Europe, Rimsting / Germany
Photo:	© Herbert Schlosser/DIGI-TEL, Oggau / Austria

05 | Square roofing, Zeltweg / Austria

Location:	Zeltweg, Styria / Austria
Completion:	2004
Roofed area:	408 m²
Membrane material type:	Glass/PTFE fabric
Client:	Verein Zeltweg Attraktiv
Architect:	Fabi & Krakau Architekten, Regensburg / Germany
General contractor for membrane roof:	Covertex, Obing / Germany
Structural engineer:	Teschner, Kosel / Germany
Design of patterning:	Teschner, Kosel / Germany
Steel erection:	Sgardelli, Knittelfeld / Austria
Membrane fabrication:	KfM, Wallhausen / Germany
Erection of membrane and wire ropes:	Montageservice SL, Hallbergmoos / Germany
Photo:	© Michael Seidel, Vienna / Austria

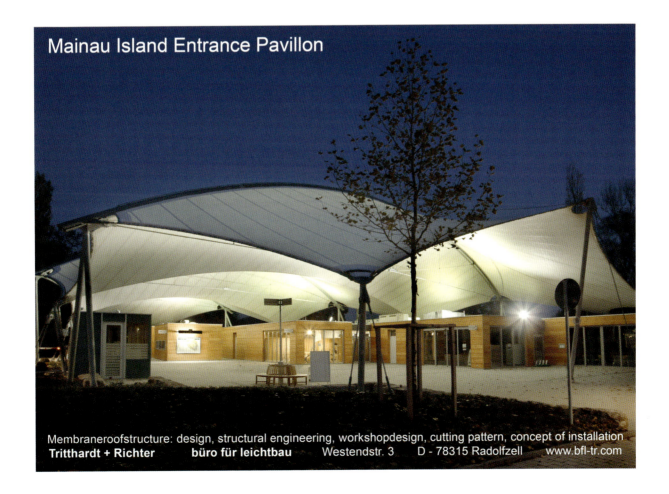

Mainau Island Entrance Pavillon

Membraneroofstructure: design, structural engineering, workshopdesign, cutting pattern, concept of installation

Tritthardt + Richter büro für leichtbau   Westendstr. 3   D - 78315 Radolfzell   www.bfl-tr.com

06 | Vehicle park roofing Waste Management Office, Munich / Germany

Location:	Munich / Germany
Completion:	1999
Roofed area:	8,400 m²
Membrane material type:	Glass/PTFE fabric
Client:	state capital Munich
Architect:	Ackermann & Partner, München / Germany
Structural engineer:	Schlaich Bergermann & Partner, Stuttgart / Germany
Design of patterning:	Tensys, Bath / United Kingdom
Erection planning:	Schlaich Bergermann & Partner
Steel erection:	Seele, Gersthofen / Germany
Wire ropes:	PFEIFER, Memmingen / Germany
Fabricator:	Birdair Europe Stromeyer, Konstanz / Germany
Membrane erection:	Birdair Europe Stromeyer, Konstanz / Germany
Photo:	© Michael Seidel, Vienna / Austria

PFEIFER

Die Leichtigkeit des Bauens

Weltweit verlassen sich Ingenieure und Architekten bei der Planung, Fertigung und Montage von Seilbauwerken auf unsere Kompetenz.

**PFEIFER
SEIL- UND HEBETECHNIK GMBH**

DR.-KARL-LENZ-STR. 66
D-87700 MEMMINGEN
TELEFON 08331-937-285
TELEFAX 08331-937-350
E-MAIL cablestructures@pfeifer.de
INTERNET www.pfeifer.de

07 | Umbrellas IHK Würzburg / Germany

Location:	Würzburg / Germany
Completion:	2003
Roofed area:	approx. 475 m²
Membrane material type:	ETFE foil, rope-supported, printed
Client:	IHK-Würzburg-Schweinfurt / Germany
Architect:	Franz Göger/Georg Redelbach, Marktheidenfeld / Germany
General contractor membrane:	Covertex, Obing / Germany
Structural engineer:	SMP Schöne/Maatz + Partner (IPZ, Berlin)
Membrane fabrication:	KfM, Wallhausen / Germany
Wire ropes:	Görlitzer Draht- und Hanfseilfabrik, Görlitz / Germany
Erection of membrane and wire ropes:	Montageservice SL, Hallbergmoos / Germany
Photo:	© Covertex, Obing / Germany

08 | Roofing of courtyard, Vienna city hall / Austria

Location:	Vienna inner city / Austria
Completion:	2000
Roofed area:	1,050 m²
Membrane material type:	PES/PVC fabric
Client:	Magistrat der Stadt Vienna
Architect:	Silja Tillner, Vienna / Austria
Structural engineer:	Schlaich Bergermann & Partner, Stuttgart / Germany
Design of patterning:	Tensys, Bath / United Kingdom
Erection planning:	Schlaich Bergermann & Partner, Stuttgart / Germany
Project management:	Vasko & Partner, Vienna / Austria
Steel erection:	Filzamer, Vienna / Austria
Wire ropes:	Augsburger Drahtseilfabrik, Friedberg / Derching / Germany
Membrane fabrication:	Spandome Center Kft, Budapest / Hungary
Membrane erection:	Covertex, Obing / Germany
Photo:	© Michael Seidel, Vienna / Austria

09 | Atrium roofing, Schöllerbank, Vienna / Austria

Location:	Vienna inner city
Completion:	2000
Roofed area:	approx. 270 m²
Membrane material type:	three-layer ETFE pillow
Client:	Schöllerbank
Architect:	Jabornegg Pálffy, Vienna / Austria
Structural engineer:	Karlheinz Wagner, Vienna / Austria
Steelwork and wire ropes:	Ma-Tec, Neutal / Austria
Fabricator of pillow:	Skyspan, Rimsting / Germany
Erection of pillow:	Montageservice SL, Hallbergmoos / Germany
Photo:	© Werner Kaligofsky, Vienna / Austria

10 | World cup football globe

Location:	Cologne / Germany
Completion:	2003
Roof area:	736 m²
Membrane material type:	printed three-layer ETFE-PVC pneumatic pillow
Client:	DFB
Architect:	Art Event, Vienna / Austria
General contractor membrane:	Covertex, Obing / Germany
Structural engineer:	Mero, Würzburg / Germany
Design of patterning:	Covertex, Obing / Germany
Material specialist report, approval of individual cases:	Labor Blum, Stuttgart / Germany
Steel erection:	Mero, Würzburg / Germany
Fabricator of pillows:	KfM, Wallhausen / Germany
Erection of pillow:	Montageservice SL, Hallbergmoos / Germany
Photo:	© Frank Rümmele, Alfter / Germany

11 | Pillow roofing, Galets Expo 2002

Location:	Neuchâtel / Switzerland
Completion:	2002
Roofed area (Galets 1, 2, 3):	approx. 14,900 m²
Membrane material type:	PES/PVC fabric pillow
Client:	Expo 02
Architect:	Multipack, Neuchâtel / Switzerland
Structural design of pillows:	IF-Group, Reichenau / Germany
Detailed design patterning of pillows:	IF-Group, Reichenau / Germany
Detailed design of steelwork:	Zwahlen & Mayr, Aigle / Switzerland
Erection planning pillow:	IF-Group, Reichenau / Germany
Material supervision and monitoring – air pressure, tension and temperature:	Labor Blum, Stuttgart / Germany
Steel erection:	Zwahlen & Mayr, Aigle / Switzerland
Membrane fabrication:	CANOBBIO, Castelnuovo Scrivia / Italy
Erection of pillow:	Inauen-Schätti, Schwanden / Switzerland
Photo:	© CANOBBIO, Castelnuovo Scrivia / Italy

12 | **Christo and Jeanne-Claude: Wrapped Reichstag, Berlin 1971–95. © Christo 1995**

Location:	Platz der Republik, Berlin / Germany
Completion:	1995
Wrapped surface:	approx. 50,000 m²
Membrane material type:	PP fabric, aluminium steamed (approx. 100,000 m²)
Client:	Verhüllte Reichstag GmbH, directors: Roland Specker und Wolfgang Volz
Concept and design:	Christo und Jeanne-Claude Javacheff
Structural design and patterning:	IPL, Radolfzell / Germany
Steel erection:	Steel erection Zwickau, Zwickau / Germany
Fabrication of wire ropes:	Gleistein Tauwerk, Bremen / Germany
Membrane fabrication:	Spreewald Planen, Vetschau / Germany; Zeltaplan, Taucha / Germany; Canobbio, Serravalle Scrivia / Italy
Fabricator of air pillows:	Heba, Emsdetten / Germany
Membrane erection:	Reichstagverhüllungsmontage GmbH
Photo:	© Wolfgang Volz, Düsseldorf / Germany

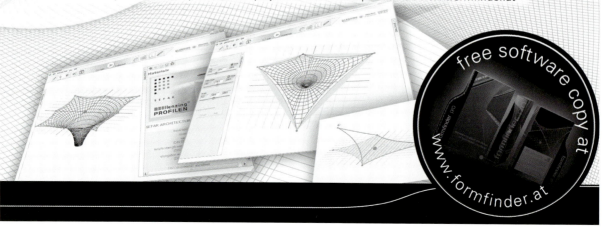

formfinder

Software to assists the design and planning of form-active structures

Formfinder assists architects and project planners in the design, planning and cost-effectiveness assessment for the implementation of form-active structures. The easy to use tool answers questions from the form-finding process up to material decisions. Advanced features like the "one click" cost estimation generates a productive dialogue between the client, the architect, the structural planner and the manufacturer of the materials used. Formfinder makes it possible to compare the desired design with projects that have already been implemented. As a result, Formfinder is the most advanced "Online search engine for form-active structures". Knowing what has already been built, when, by whom, where and how is of considerable commercial value and forms the basis for every successful implementation project. To learn more please visit www.formfinder.at

13 | Concert awning, Radolfzell / Germany

Location:	Radolfzell bank promenade
Completion:	1989
Roofed area:	340 m²
Material:	rope net with covering of polycarbonate boards
Client:	Stadt Radolfzell / Germany
Architect:	IPL, Radolfzell / Germany
Structural design and patterning:	IPL, Radolfzell / Germany
Erection planning:	IPL, Radolfzell / Germany
Steel erection:	Späth, Steißlingen / Germany
Wire ropes:	PFEIFER, Memmingen / Germany
Fabrication of boards:	Carl Nolte, Greven / Germany
Erection of rope net:	Schätti, Tschuggen / Switzerland
Photo:	© Michael Seidel, Vienna / Austria

14 | Roofing Rhönklinikum / Germany

Location:	Bad Neustadt/Saale / Germany
Completion:	1998
Roofed area:	approx. 1,200 m²
Material:	rope net, glass covering
Client:	Rhön-Klinikum AG, Bad Neustadt / Germany
Architect:	Lamm, Weber, Donath & Partner, Stuttgart / Germany
Structural engineer:	Werner Sobek Ingenieure, Stuttgart / Germany
Glass tile system:	Werner Sobek Ingenieure, Stuttgart / Germany
Detailed design of cutting patterns for rope net:	IF-Group, Reichenau / Germany
Erection planing of rope net:	IF-Group, Reichenau / Germany
Steel erection:	Mero, Würzburg / Germany
Ropes and rope net:	Brugg (Fatzer AG), Romanshorn / Switzerland
Erection of rope net:	Inauen-Schätti, Tschachen / Switzerland
Photo:	© MERO-TSK International, Würzburg / Germany

15 | Pillow roofing, Tropical Island / Germany

Location:	Brand / Germany
Completion:	2005
Roofed area:	approx. 20,000 m²
Membrane material type:	three-layer ETFE large pillow
Client:	Tropical Islands Asset Management, Krausnick / Germany
Architect:	CL Map, Munich / Germany
General contractor for membrane roof:	CENO TEC, Greven / Germany
Structural engineer:	form TL, Radolfzell / Germany
Design of patterning:	CENO TEC, Greven / Germany; form TL, Radolfzell / Germany
Erection planning:	Montageservice SL, Hallbergmoos / Germany
Steel erection:	Thyrolf & Uhle, Dessau / Germany
Wire ropes:	Berndorf FAS, Berndorf / Germany
Membrane fabrication:	CENO TEC, Greven / Germany
Erection of pillows, rope nets, secondary steelwork and gutters:	Montageservice SL, Hallbergmoos / Germany
Photo:	© CENO TEC, Greven / Germany

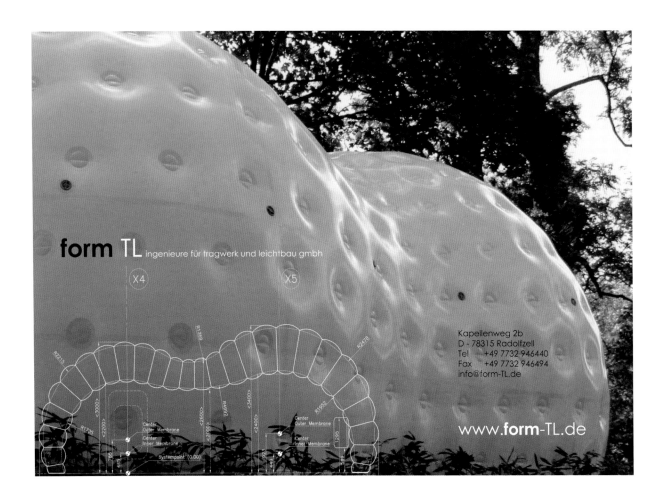

16 | Roofing of Airport Center MAC, Munich / Germany

Location:	Munich airport / Germany
Completion:	1999
Roofed area:	approx. 7,500 m²
Membrane material type:	Glass/PTFE fabric
Client:	Flughafen München / Germany
Architect:	Murphy/Jahn, Chicago / USA
Structural engineer:	Ove Arup
Steel erection:	Steel erection Wolf, Rosenheim / Germany
Supply of wire ropes:	Redaelli, Cologno Monzese / Italy
Membrane erection:	Montageservice SL, Hallbergmoos / Germany
Photo:	© Engelhardt / Sellin, Aschau / Germany

17 | Roofing of access building, Suvarnabhumi International Airport Bangkok / Thailand

Location:	Samut Prakarn Province, Bangkok / Thailand
Completion:	2005
Area:	approx. 108,000 m²
Membrane material type (three-layer):	Outer: Glass/PTFE fabric; Middle: Polycarbonate boards on rope net; Inner: glass fibre fabric, PTFE-Ter-Polymer-coated and aluminium steamed (Low-E coated)
Client:	Royal Thai Government / Thailand
General contractor:	ITO Joint Venture
Architect:	Murphy/Jahn, Chicago / USA; ACT Consultants, Bangkok / Thailand; TAMS Consultants/Earth Tech, New York / USA
Structural engineer:	Werner Sobek Ingenieure, Stuttgart / Germany
General contractor for membrane roof:	Hightex, Rimsting / Germany
Design of patterning:	Tensys, Bath / United Kingdom
Acoustics/material development and Material supervision:	Labor Blum, Stuttgart / Germany
Air conditioning:	Transsolar Energietechnik, Stuttgart / Germany
Coating Low-E membrane:	Polymade ITT, Bergheim / Germany
Glass fibre fabric:	Interglas, Erbach / Germany
Membrane fabrication:	Hightex
Membrane erection:	Hightex
Photo:	© Rainer Viertlböck, Architecture Photographie

18 | **Façade of Vienna airport control tower / Austria**

Location:	Vienna Schwechat / Austria
Completion:	2005
Area:	3,324 m²
Membrane material type:	Glass/PTFE fabric
Client:	Flughafen Vienna / Austria
Architect:	Zechner & Zechner, Vienna / Austria
General contractor membrane:	Covertex, Obing / Germany
Structural engineer:	Lorenz Consult, Graz / Austria
Membrane design and form finding:	Peter Mandl, Graz / Austria
Design of patterning:	form TL, Radolfzell / Germany
Materials specialist report:	Labor Blum, Stuttgart / Germany
Erection planning steel erection:	Zeman, Vienna / Austria
Steel erection und Montage:	Zeman, Vienna / Austria
Membrane fabrication:	KfM, Wallhausen / Germany
Membrane erection:	Montageservice SL, Hallbergmoos / Germany
Photo:	© Herbert Schlosser/DIGI-TEL, Oggau / Germany

your partner for special installation – www.montage-service.com
Montageservice SL GmbH ▪ Ismaninger Straße 98 ▪ 85399 Hallbergmoos ▪ Germany ▪ info@montage-service.com

MONTAGESERVICE SL GmbH

19 | Roofing of Fröttmaning station, Munich / Germany

Location:	Fröttmaning, Munich / Germany
Completion:	2005
Roofed area:	approx. 7,200 m²
Membrane material type:	Glass/PTFE fabric
Client:	Landeshauptstadt Munich, Baureferat
Architect:	Bohn Architekten, Munich / Germany
General contractor for membrane roof:	Covertex, Obing / Germany
Structural engineer:	Christoph Ackermann, Munich / Germany
Design of patterning:	Teschner, Kosel / Germany
Compensation tests and approval of individual cases:	Labor Blum, Stuttgart / Germany
Steel erection:	Maurer Söhne, Munich / Germany
Wire ropes:	PFEIFER, Memmingen / Germany
Membrane fabrication:	KfM, Wallhausen / Germany
Membrane erection and Wire ropes:	Montageservice SL, Hallbergmoos / Germany
Photo:	© Florian Scheiber, Munich / Germany

20 | Roofing of Dresden Main Station / Germany

Location:	Main railway station, Dresden / Germany
Completion:	2006
Roofed area:	approx. 30,000 m²
Membrane material type:	Glass/PTFE fabric
Client:	Deutsche Bahn AG
Architect:	Foster & Partners, London / England
Structural engineer:	Happold, Berlin and London / Germany and England
Membrane fabrication:	Skyspan Europe, Rimsting / Germany
Membrane erection:	Montageservice SL, Hallbergmoos / Germany
Photo:	© Michael Seidel, Vienna / Austria

21 | Roofing of stands, Formula 1 Ring, Istanbul / Turkey

Location:	Istanbul / Turkey
Completion:	2005
Roofed area:	approx. 18,000 m²
Membrane material type:	PES/PVC fabric
Client:	FIYAS Formular Istanbul Yatirim A.S., Istanbul / Turkey
Architect:	ORION Insaat & Dekorasyon, Istanbul / Turkey; Teschner, Kosel / Germany
General contractor for membrane roof:	CENO TEC, Greven / Germany
Structural engineer and Design of patterning:	Teschner, Kosel / Germany
Erection planning:	Montageservice SL, Hallbergmoos / Germany
Steel erection:	Evren, Istanbul / Turkey
Wire ropes:	Görlitzer Draht- und Hanfseilfabrik, Görlitz / Germany; PFEIFER, Memmingen / Germany
Membrane fabrication:	CENO TEC, Greven / Germany
Membrane erection:	Montageservice SL, Hallbergmoos / Germany
Photo:	© CENO TEC, Greven / Germany

22 | Roofing of stand, Estádio Intermunicipal, Faro-Loulé / Portugal

Location:	Faro / Portugal
Completion:	2003
Roofed area:	10,168 m²
Membrane material type:	PES/PVC fabric (precontraint)
Client:	Town council of Faro und Loulé / Portugal
Architect:	HOK Sport, London / England; AARQ Atelier de Arquitectura, Lissabon / Portugal; W.S. Atkins, London / England
Structural engineer:	W.S. Atkins, London / England; Tensys, Bath / England
General contractor for membrane roof:	Somague, Sintra / Portugal
Design of patterning:	IPL, Radolfzell / Germany
Erection planning wire ropes:	Schlaich Bergermann & Partner, Stuttgart / Germany
Steel erection:	SIMI, Lisbon / Portugal
Wire ropes:	PFEIFER, Memmingen / Germany
Erection of cable structure:	PFEIFER, Memmingen / Germany
Membrane fabrication:	CENO TEC, Greven / Germany
Membrane erection:	Montageservice SL, Hallbergmoos / Germany
Photo:	© CENO TEC, Greven / Germany

23 | Roofing of stands, Volkswagen Arena / Germany

Location:	Allerpark, Wolfsburg / Germany
Completion:	2002
Roofed area:	approx. 15,000 m²
Membrane material type:	PES/PVC fabric
Client:	Wolfsburg AG, Wolfsburg / Germany
Architect:	Büro Hpp, Düsseldorf / Germany
General contractor for membrane roof:	CENO TEC, Greven / Germany
Structural engineer:	Schlaich Bergermann & Partner, Stuttgart / Germany
Design of patterning:	Teschner, Kosel / Germany
Erection planning:	Schlaich Bergermann & Partner, Stuttgart / Germany; CENO TEC, Greven / Germany
Steel erection:	dbn–Planungsgruppe Dröge-Baade-Nagaraj, Salzgitter / Germany
Wire ropes:	PFEIFER, Memmingen / Germany
Membrane fabrication:	CENO TEC, Greven / Germany
Membrane erection:	CENO TEC, Greven; Montageservice SL, Hallbergmoos / Germany
Photo:	© CENO TEC, Greven / Germany

24 | Stand roofing, Fenerbahce stadium / Turkey

Location:	Istanbul / Turkey
Completion:	2002
Roofed area:	approx. 20,400 m²
Membrane material type:	PES/PVC fabric
Client:	Fenerbahce, Istanbul / Turkey
Architect:	Teschner, Kosel / Germany (Membrandach); A & Z Akzu, Istanbul / Turkey (Tribünen)
General contractor for membrane roof:	CENO TEC, Greven / Germany
Structural engineer und design of patterning:	Teschner, Kosel / Germany
Compensation tests:	Labor Blum, Stuttgart / Germany
Erection planning:	Montageservice SL, Hallbergmoos / Germany
Steel erection:	Temsan, Ankara / Turkey
Membrane fabrication:	CENO TEC, Greven / Germany
Membrane erection:	Montageservice SL, Hallbergmoos / Germany
Photo:	© CENO TEC, Greven / Germany

25 | Stand roofing Sheikh Khalifa Bin Zayed Stadium / United Arab Emirates

Location:	Al Ain / United Arab Emirates
Completion:	2004
Roofed area:	approx. 7,500 m²
Membrane material type:	Glass/PTFE fabric
Client:	Government of Abu Dhabi / United Arab Emirates
Architect:	Rice Perris Elli, Dubai / United Arab Emirates; Crang & Boake, Toronto / Canada
General contractor for membrane roof:	Mero, Würzburg / Germany
Structural engineer:	Mero, Würzburg / Germany
Design of patterning:	If-Group, Reichenau / Germany
Erection planning:	CENO TEC, Greven / Germany; If-Group, Reichenau / Germany
Steelwork:	Mero, Würzburg / Germany
Secondary steel:	Montageservice SL, Hallbergmoos / Germany
Membrane fabrication:	CENO TEC, Greven / Germany
Membrane erection:	Montageservice SL, Hallbergmoos / Germany
Erection of wire ropes:	Montageservice SL, Hallbergmoos / Germany
Photo:	© MERO-TSK International, Würzburg / Germany

26 | Roofing of Arénes de Nîmes / France

Location:	Nîmes / France
Completion:	1988
Roofed area:	approx. 4,000 m²
Membrane material type:	rope-supported PES/PVC fabric pillow
Client:	Ville de Nîmes / France
Architect:	Finn Geipel & Nicolas Michelin, Paris / France
Structural engineer:	Schlaich Bergermann & Partner, Stuttgart / Germany (Rudolf Bergermann, Werner Sobek, Jochen Bettermann)
Design of patterning:	Schlaich Bergermann & Partner, Stuttgart / Germany
Erection planning:	Schlaich Bergermann & Partner, Stuttgart / Germany; IF-Group, Horst Dürr, Konstanz / Germany
Steel erection:	Baudin Chateauneuf, Nîmes / France
Membrane fabrication:	Stromeyer, Konstanz / Germany
Textile Acoustic module:	Koch Hightex, Rimsting / Germany
Photo:	© PFEIFER, Memmingen / Germany

27 | Roofing of stands, Estadio Olimpico, Seville / Spain

Location:	Seville / Spain
Completion:	1999
Roofed area:	approx. 25,000 m²
Membrane material type:	PES/PVC fabric
Client:	Sociedad Estadio Olimpico de Seville / Spain; Junta de Andaluciá
Architect:	Cruz `Y Ortiz, Seville / Spain
General contractor for membrane roof:	ACS, Seville / Spain
Structural engineer and Design of patterning:	Schlaich Bergermann & Partner, Stuttgart / Germany
Erection planning:	Schlaich Bergermann & Partner, Stuttgart / Germany
Steel erection:	ACS, Seville / Spain
Supply and erection of cable structure:	PFEIFER, Memmingen / Germany
Membrane fabrication:	CENO TEC, Greven / Germany
Membrane erection:	Montageservice SL, Hallbergmoos / Germany
Photo:	© CENO TEC, Greven / Germany

28 | Roofing of Waldstadion, Frankfurt / Germany

Location:	Frankfurt am Main / Germany
Completion:	2005
Roofed area:	approx. 27,000 m²
Adaptable part:	approx. 8,000 m²
Membrane material type:	Glass/PTFE fabric
Adaptable part:	PES/PVC fabric
Client:	Waldstadion Frankfurt am Main / Germany, Gesellschaft für Projektentwicklung mbH
Architect:	gmp – von Gerkan, Marg and Partners
Structural engineer:	Schlaich Bergermann & Partner, Stuttgart / Germany
Design of patterning:	IF-Group, Reichenau / Germany
Approval of individual cases:	Labor Blum, Stuttgart / Germany
Erection planning:	Schlaich Bergermann & Partner, Stuttgart / Germany
Erection planning membrane:	IF-Group, Reichenau / Germany
Steel erection:	Max Bögl, Neumarkt / Germany
Wire ropes:	PFEIFER, Memmingen / Germany; Augsburger Drahtseilfabrik, Friedberg / Derching / Germany
Membrane fabrication:	Skyspan Europe, Rimsting / Germany
Erection of cable structure:	Inauen-Schätti, Tschachen / Switzerland; VSL, Subingen / Switzerland
Membrane erection:	Skyspan Europe, Rimsting / Germany
Photo:	© Heiner Leiska, Kiel / Germany

29 | Roofing of stands, Gottlieb Daimler Stadion / Germany

Location:	Stuttgart / Germany
Completion:	1992
Roofed area:	approx. 34,000 m²
Membrane material type:	PES/PVC fabric
Client:	Sportamt/Hochbauamt Stuttgart / Germany
Architect:	Weidleplan Consulting, Stuttgart / Germany; Siegel & Partner, Stuttgart / Germany
Design of patterning:	Schlaich Bergermann & Partner, Stuttgart / Germany
Design of patterning:	Schlaich Bergermann & Partner, Stuttgart / Germany
Erection planning:	Schlaich Bergermann & Partner, Stuttgart / Germany
Steel erection:	Haslinger, München / Germany
Cable structure:	PFEIFER, Memmingen / Germany
Membrane fabrication:	Koch Hightex, Rimsting / Germany
Membrane erection:	Koch Hightex, Rimsting / Germany
Photo:	© PFEIFER, Memmingen / Germany

30 | Roofing of Bullfight Arena, Vista Alegre, Madrid / Spain

Location:	Calle de Nueva Contrucción, Madrid / Spain
Completion:	2000
Roofed area:	approx. 1,960 m²
Material type, pillow:	PES/fabric (outer), Rope-supported ETFE foil (inner)
Client:	Arturo Beltrán, Palumi S.A., Madrid / Spain
Concept and structural engineer:	Schlaich Bergermann & Partner, Stuttgart / Germany
Design of patterning:	Schlaich Bergermann & Partner, Stuttgart / Germany
Erection planning:	Schlaich Bergermann & Partner, Stuttgart / Germany
Wire ropes:	PFEIFER, Memmingen / Germany
Membrane fabrication:	Skyspan, Rimsting / Germany
Membrane erection:	Skyspan, Rimsting / Germany
Photo:	© Roland Halbe Fotografie, Stuttgart / Germany

31 | Roofing of Velodrome (redesign), Abuja / Nigeria

Location:	Abuja, Nigeria
Completion:	2006
Roofed area:	10,649 m²
Membrane material type:	Glass/PTFE fabric
Client:	Federal Government of the Federal Republic of Nigeria
General contractor:	Bilfinger Berger AG, Wiesbaden / Germany, Julius Berger Nigeria Plc., Abuja / Nigeria
Structural engineer:	form TL, Radolfzell / Germany
Design of patterning:	form TL, Radolfzell / Germany
Materials testing:	Labor Blum, Stuttgart / Germany
Erection planning:	IF-Group, Reichenau / Germany
Steel erection:	PFEIFER, Memmingen / Germany
Wire ropes:	PFEIFER, Memmingen / Germany
Erection of rope net:	PFEIFER, Memmingen / Germany
Membrane fabrication:	CANOBBIO, Castelnuovo Scrivia / Italy
Membrane erection:	Montageservice SL, Hallbergmoos / Germany
Photo:	© Bilfinger Berger Nigeria GmbH, Wiesbaden / Germany

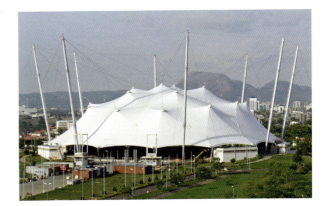

32 | Roofing of stands, Jaber Al-Ahmad Stadium/Kuwait

Location:	Kuwait City / Kuwait
Completion:	2007
Roofed area: approx.	approx. 40,000 m²
Membrane material type:	Glass / PTFE fabric
Client:	Ministry of Public Works, Kuwait
Structural engineer:	Weidleplan Consulting, Stuttgart / Germany
Structural engineer:	Schlaich Bergermann & Partner, Stuttgart / Germany
Consulting for ring beam, membrane, wire ropes:	IF-Group, Reichenau / Germany (for Weidleplan)
Design of patterning:	Birdair, Amherst / USA
Wire ropes:	BRIDON/BTS Drahtseile, Doncaster and Gelsenkirchen / Germany
Erection of wire ropes:	Montageservice SL, Hallbergmoos; PFEIFER, Memmingen / Germany
Membrane fabrication:	Birdair, Amherst / USA
Membrane erection:	Birdair, Amherst / USA; Montageservice SL, Hallbergmoos / Germany
Photo:	© Montageservice SL, Hallbergmoos / Germany